Blazing the Trail: Essays by Leading Women in Science

Emma Ideal

Rhiannon Meharchand

To all those devoted to advancing the store of human knowledge: the trailblazers, for their courage and inspiration; today's scientists, for their stewardship and leadership; and the next generation of explorers, whose efforts will enable discovery beyond our imagination.

CONTENTS

	Preface	viii
	Acknowledgments	xi
1	Christine A. Aidala	1
2	Susan Davis Allen	9
3	Ani Aprahamian	19
4	Sheila Brown Bailey	27
5	Olgica Bakajin	37
6	Daniela Bortoletto	46
7	Patricia R. Burchat	53
8	Cathryn Carson	59
9	Shirley Chiang	66
10	Janet Conrad	75
11	Esther Conwell	81
12	Jill P. Dahlburg	87
13	Arati Dasgupta	94
14	Sarah M. Demers	101
15	Mildred Dresselhaus	105
16	Lucy Fortson	114

17   Elsa Garmire   123

18   Jarita C. Holbrook   135

19   Carolina C. Ilie   143

20   Barbara A. Jones   150

21   Noémie Benczer Koller   158

22   Lillian Christie McDermott   166

23   Anne-Marie Novo-Gradac   177

24   Linda J. Olafsen   183

25   Angela V. Olinto   189

26   Marjorie A. Olmstead   197

27   Michelle L. Povinelli   206

28   Diane (Betsy) Pugel   213

29   Juana Rudati   218

30   Michelle D. Shinn   224

31   Elizabeth H. Simmons   230

32   Elma Beth Snipes   236

33   Meg Urry   244

34   Lakeisha Maria Hogue Walker   252

35   Alice E. White   258

      About The Editors   265

PREFACE

In April 2011, we flew halfway across the world to attend the 4th IUPAP International Conference on Women in Physics (ICWIP) in Stellenbosch, South Africa. The ICWIP conference included a series of workshops, one of which was focused on methods for attracting girls to physics. There, Ram Ramaswamy presented a book, <u>Lilavati's Daughters: The Women Scientists of India</u>, containing 90 short essays by female Indian physicists. The goal: provide a suite of highly successful physicists to serve as modern-day role models for young women in India. Inspired, we came home and immediately started working on an analogous text for an American audience. This book is the culmination of that work.

The purpose of this book is threefold: to provide a diverse group of role models; to inform young people about scientific careers (physics in particular); and to encourage students to study science and pursue scientific careers. For role models, we assembled a diverse group of women who come from a wide variety of backgrounds but share a passion for science and discovery. To make sure that as many young women as possible could find someone to identify with and look up to, essayists were instructed to include information about themselves and their formative years including education, hobbies, and personal life. To inform readers about scientific careers, we solicited essays from women who have achieved great success in a wide variety of careers, in the academic, industry, government, and other sectors. We encouraged them to share how they came to study science and what steered them to their current path. Finally, to inspire readers to pursue scientific studies and careers, we asked essayists to not only share career advice but also reflect on their experiences – both positive and negative – candidly sharing challenging times but focusing whenever possible on lessons learned.

RHIANNON MEHARCHAND

I could have used a book like this. Growing up, I had no idea who scientists were, what they did, or how to become one. Yet when the aptitude test said "theoretical/mathematical physicist", with the encouragement of my high-school physics teacher, I went all-in for my college applications and declared a Physics major. The first in my immediate family to attend college, I stumbled clumsily through my studies, leaning heavily on academic advisors and word-of-mouth to figure out what to do next. Thanks to luck, hard work, and several summer research programs I got into a great graduate program, where an extremely supportive mentor, countless opportunities for scientific and professional growth, more luck and more hard work allowed me to add "Ph.D." to my list of family firsts. Now a postdoc at a national laboratory, I'm still finding my way – seeking wisdom from those who've gone before and reaching for the next step in a (hopefully) long and successful career.

Role models are important. They expand horizons, opening eyes to new opportunities. They provide guidance, shepherding one along the path to success. They advise, sharing lessons learned along their personal journey. I hope this book provides real-life role models for young women everywhere, and inspires them to dream big, work hard, and achieve things they never thought possible.

I feel blessed to have had many opportunities to work with smart, talented people – most notably Emma Ideal, whose initiative and drive brought this book to life. Emma, I have truly enjoyed working with you and know this is just a blip on what will be a long list of achievements in your career. I'm honored to have been able to collaborate with you, and conclude this project considering you a close colleague and friend.

EMMA IDEAL

My trajectory through physics, and the academic sphere more broadly, has been profoundly influenced by those teachers, mentors, and role models who saw my potential and encouraged me to aim high. At each stage in my academic growth, I have been fortunate enough to have such a support structure; it has kept me motivated to dream big, and when faced with tough challenges, to keep my chin up and carry on! An illustrative—not to mention, relevant—example is this book. When a second-year Ph.D. student approaches her advisor to say she'd like to "make a book", you might expect the terse reply: "No, there's no time for that" or "HA! Oh…you're serious? Okay, good luck." Rather, I received two enthusiastically supportive replies: "Yes absolutely, let's do it. Now, for publishing options, I think…"

However, I don't chalk this all up to luck! You can—and ought to—be actively seeking mentors who believe in you, have your best interest in mind, and inspire you to aim high. To advance this process, it is my hope that this book will provide you with 35 new role models whose words will resonate with you, exposing you to new ideas and opportunities and preparing you for your successful career ahead. Onward!

I finally add a few personal sentiments. To my mentors that have always been there: my father Lee Ideal, for being a steadfast source of strength to me in all my endeavors, and my mother Elvira Ideal, who by example teaches me to be a giver—the most important quality one can develop in life. To Rhiannon Meharchand, my trusty collaborator on <u>Blazing the Trail</u>: it has been a true pleasure completing this volume with you; I look up to you as a mentor and am lucky to call you a close friend. Last but never least, to my brother Paul Ideal; you will always be my hero.

ACKNOWLEDGMENTS

There are many people who were instrumental in this endeavor, and we would like to take this opportunity to give them our warmest thanks. First and foremost, we thank Professor Ram Ramaswamy, who provided the inspiration for this project in April 2011 when he presented on his and Professor Rohini Godbole's book, <u>Lilavati's Daughters</u>, at the 4th IUPAP International Conference on Women in Physics (ICWIP) in Stellenbosch, South Africa. We would also like to thank the ICWIP organizers, the U.S. delegation leaders, the National Superconducting Cyclotron Laboratory (NSCL) at Michigan State University (MSU) and Yale University for providing two young, impressionable graduate students the transformative experience of attending ICWIP.

Professors Meg Urry and Sarah Demers – both featured essayists – provided guidance and support from the project's onset to its completion. Dr. Carol Guess graciously spent her spare time copy-editing final drafts of each and every essay. Ms. Ariel Ekblaw, our artistic connection, helped us discover Ms. Jane Long, who designed the cover. Without Jane, who knows what two artistically-challenged people would have come up with…

A huge thanks is extended to the Yale University Physics Department and NSCL/MSU, for providing funding for purchasing and disseminating copies of this book.

Last but not least, we are deeply indebted to all our inspiring essayists for their time and generous contributions, for without their hard work, there would be no book.

CHRISTINE A. AIDALA

Bona Fides

Dr. Christine A. Aidala received her Bachelor's degree in Physics and Music from Yale University in 1999 and her Ph.D. in Physics from Columbia University in 2005. She is currently transitioning from a staff scientist position with Los Alamos National Laboratory to a tenure-track position in the Physics Department at the University of Michigan. Her recognitions include being a Distinguished Woman Physicist Colloquium speaker at the University of Connecticut, a recipient of the Sambamurti Memorial Award from Brookhaven National Laboratory, and a recipient of the Luise Meyer-Schutzmeister Award from the Association for Women in Science. She works in experimental high-energy nuclear physics, at the border between nuclear and particle physics. She presently researches nucleon structure and quantum chromodynamics, the theory of the strong force, as part of two relatively large international collaborations, working on the PHENIX experiment at the Relativistic Heavy Ion Collider at Brookhaven National Laboratory as well as the SeaQuest experiment at Fermilab.

The Foundations of a Physicist

I'm lucky enough to be the daughter of two lifelong learners, and I can hardly think of a more valuable gift to give someone than a love of learning. My mother received her Master's degree in accounting when I was a baby, and as a teenager I was proud and inspired to witness her go back later in life and get her CPA. My father is an aerospace engineer, and growing up my sister and I sat through many a dinner-table science lesson. Our budding intellectual curiosity was sometimes rewarded with longer-winded answers than we'd bargained for, but our expanding horizons never failed to lead us to new questions. Our parents cultivated diverse interests in us, many of which continue to play a role in our lives; however, there was evidently something special about physics, as my sister ended up a physicist as well!

Perhaps in contrast to many, one of the main reasons I was convinced that I was interested in physics as I went off to college was because I had found that I liked physics *despite* my high school teacher! Giving credit where it's due, however, even though I felt that I had a poor high school physics experience I sincerely appreciated the fact that my teacher recognized my abilities. Thanks to his nomination, at the end of high school I received an award from the American Nuclear Society.

Undergraduate Endeavors

I had an inspirational and highly challenging physics class my freshman year at college, taking the fifth-level intro course, "Intensive Introductory Physics." The class started with 35 students, and 18 of us stuck it out through the end of the second semester. We became a pretty tight group in several ways in spite of our diverse backgrounds. One of the things I never would have anticipated was heading to the gym every Wednesday after handing in our legendary problem sets to blow off steam playing "Physics 260 basketball" together! My sophomore year, I co-organized Friday pizza lunch lectures by faculty for undergrads (there was always good attendance by grad students as well because of the free food!). I continued the lunches through junior and senior year, and also expanded things to reinstate the local chapter of the Society of Physics Students, which had been inactive for more than a decade.

2

Rounding it all out, I took the initiative to revive the annual spring departmental picnic. Years later, after I'd finished my Ph.D., I was talking about where I did my undergrad, and someone asked me, "Wasn't the physics department very cold there?" I was taken aback by the question, since such a description never would have occurred to me. However, thinking back over my time there, it occurred to me that it probably was – or rather would have been – except that I made it otherwise. Organizing on the order of 20 talks by faculty per year over three years, I had interacted with just about every professor in the department. I had developed a very close relationship with the chair, who kindly provided financial support for the pizza talks and willingly expanded his support for things like the renewed department picnics when I offered to organize them. Of course, organizing so many activities within the department, I had also developed relationships with plenty of my fellow students.

It was very valuable to have sought out research opportunities throughout undergrad. I first worked in a lab the summer after my freshman year, at my home university. I applied to and attended summer research programs in different places each year after that, which were all well worthwhile. They gave me a taste of what a career in research would be, which bore very little resemblance to undergraduate lecture or lab courses. Spending each summer in a new place also exposed me to a variety of research projects and environments and certainly expanded my scientific social network as well.

Difficult Times

I did have some tough years after undergrad. I started a Ph.D. program immediately after I finished my Bachelor's and ended up going on leave shortly after the second term had started. The initial cause was some health problems, but I had been exceedingly unhappy there and wasn't motivated to go back after missing more than a month of classes. I had encountered a number of negative circumstances, in particular

3

highly unprofessional treatment by certain faculty members. I learned the lesson that even faculty members are not beyond lying about departmental rules or exploiting a student's financial support. On top of that, I had begun working with a research group there the summer before I started. A postdoc was assigned to supervise me, and he told me flat-out, "Graduate students can't ask any questions." Looking back, it seems outrageous that I should have somehow accepted this, but he was dead serious, and if I tried to ask questions, he would turn me away, telling me to "think harder" and even to "stare at it longer." I had days when I would sit "staring" at my computer, with tears rolling down my cheeks, until the clock struck five and I felt I could finally leave, going home ashamed and depressed. So when I had health problems six months later, I didn't return to the university, and I ended up spending a year and a half in Italy, the home country of my then-boyfriend, now husband of 11 years, teaching English and music.

I had a lot of baggage as a grad school "dropout" throughout that period, carrying some of it with me well beyond. It's amazing to me to this day that my (now) husband stuck by me through all of it, given that we hadn't even been together for a year when my mental health started declining. But I slowly recovered and convinced myself that it had been the environment that wasn't right for me, not the goal of a Ph.D. in physics. I looked into doing my Ph.D. in Italy, but given the education system there at that time, they wouldn't recognize my Bachelor's degree. Redoing four years of undergrad didn't seem like a feasible option, but I knew I didn't want to return to my previous institution, and I had missed the window to try to transfer to another program in the U.S. I decided to look for a temporary research position instead.

To my disappointment, several contacts from my undergrad days and earlier research projects failed to lead to anything concrete. But perseverance eventually won out, and I was fortunate enough to find someone willing to take a chance on me based only on my resume and a phone interview. I was hired for a 12-month research job at Brookhaven National Lab (BNL) working on one of the experiments at the Relativistic Heavy Ion Collider. I was certainly grateful, but I had no inkling, at the time, how that initial job would have such a tremendous impact on the next 11 years of my career and on my career-family strategy. During that temporary research position, I decided to reapply to grad school, only considering places involved in research at the collider and that were within a couple-hour "commute" from BNL. My newlywed husband, a software engineer, had found a job at BNL himself, and we didn't want to move again after just a year. Time passed, and I ended up sitting in the same hallway at BNL: for that 12-month job, 3 ½ years after that to complete my Ph.D., a three-year postdoc stationed remotely from my employing university, a three-year named fellowship for Los Alamos National Lab (LANL), and I write this now six months into a LANL staff scientist position. I've had an office in the same hallway for 11 years, paid by four different institutions. In the meantime, my husband and I have had two children. Repeatedly negotiating to remain stationed at my large, user-facility-based experiment has offered geographic stability to my family despite the string of short-term positions typical for an early academic career.

Family Matters

I decided to start a family during the last year of my Ph.D. My personal situation allowed it – I got married in the first half of my twenties, and I figured that my Ph.D. years would be the most flexible period in my career for many years to come. With some students finishing in four years and others taking eight, what could somebody say to me if I took six months longer than I otherwise would have? In the end, I don't actually feel that I graduated any later for having had a baby, since the idea of having a baby in "the last year of my Ph.D." turned out to provide a much more meaningful deadline in my own mind than any generic notion of, "I should finish soon now." I worked so hard those last few months, and I proudly defended the day before my son's first birthday, completing all the Ph.D. requirements from scratch in only seven semesters, since the program didn't allow transfer of credits from other institutions.

Knowing I'd have limited travel opportunities once the baby was born, I arranged to give a number of talks beforehand. One opportunity led to another, and to my own amazement, I ended up giving 18 talks in the 12 months before my son was born! A handful were conference and workshop talks, but most were seminars that I arranged. Giving a number of hour-long seminars before applying to postdocs worked out wonderfully – I got a lot of practice giving interview-style talks with none of the pressure of an actual interview! Plus, it was certainly excellent for networking. The year after my son was born I continued to be proactive in seeking professional opportunities but needed to look for more that didn't involve travel. I feel that my biggest success toward that end was taking it upon myself to organize a local physics workshop, which I appreciated as a relatively rare opportunity while still a grad student.

I'd be remiss not to mention how wonderfully supportive so many colleagues were during the period after I had my son, with only one (albeit stark) exception. Working in a collaboration of >500 people, it was touching that just under 100 of my collaborators took the time to send me personal e-mail, a card, or a small gift when my son was born. My Ph.D. supervisor was extraordinarily flexible regarding my schedule, and the senior faculty member in my thesis group caught me by surprise with his genuine excitement. I recall that he spontaneously declared, "This will be a first for the Department of Energy group!" I found it an amusing and unexpected way to look at my pregnancy as a grad student, but I welcomed his unambiguous expression of support.

I had my daughter halfway through my three-year postdoc, which was definitely more challenging, in part because of the fixed-term contract, and in part because of some general tensions felt during that time. She was a more difficult baby as well, and of course at that point we had not just a baby but a baby plus a toddler. I was getting more invitations to travel, most of which I was reluctant to give up, but it was quite difficult for my husband to handle both children while I was away. It took a few months, but it finally occurred to me to try to take the little one with me. She ended up coming on six professional trips with me before she turned one. Usually I had the person who invited me help me to find local childcare, and for one workshop my mother was kind enough to take a week off from work and travel with me to watch her. It was a totally exhausting year, but we made it through, and it's gotten easier since then.

Making Choices, Settling Down

I find myself at a pivotal point in my career: about to move from my LANL staff position to a tenure-track faculty position at the University of Michigan. Working remotely for LANL, stationed at BNL, has been an excellent career opportunity for me and has been great for my family, but it was never really a long-term career option. I've known for several years that my preference would be to work at a university, with opportunities to teach and mentor junior colleagues in research. On top of that, being from relatively urban areas, and with all of my husband's family still in Europe, we felt that moving out to Los Alamos wasn't the right solution for us geographically. So I've been on the job market for faculty positions, and it's certainly been an emotional roller coaster as I've posed time and again the question, "Where could we all be happy?" I turned down my first tenure-track offer five years ago because the timing wasn't right to move my family and then an exceptionally flattering unsolicited offer the year after. Declining offers for tenure-track positions to sit in temporary ones was nerve-wracking, to say the least. I received all my offers serially, so each time a decision had to be made without knowing what opportunities might arise – or not – in the future. I made an earnest effort to pin down the matrix of personal and professional compromises I'd be willing to make, and I became much more selective about where I applied. I have great optimism that UMich and Ann Arbor will finally provide the long-term solution for which I've been searching, and it's extremely gratifying to see that years of hard work, compromise, and patience have brought me here. So it's a perfect point to take stock and reflect on how I've reached where I am and the people who've supported me so much throughout the years. I feel that true success is achieving happiness, which I know I won't fulfill unless I'm actively making both the personal and professional aspects of my life work. So, (fingers crossed!) . . . I think it's been worth the wait to settle into the right long-term position for me!

SUSAN DAVIS ALLEN

Bona Fides

Dr. Susan Davis Allen received her B.S. in Chemistry from Colorado College and Duke University, and her Ph.D. in Chemical Physics from the University of Southern California (USC). She completed a postdoctoral research appointment at USC before starting her self-proclaimed "random walk" career, which includes technical/research staff positions, extensive academic administration experience, and professorships in chemistry, physics, electrical engineering, computer science/engineering, and mechanical engineering at USC, University of Iowa, Tulane, Florida State/Florida A and M University, Arkansas State University and Embry Riddle Aeronautical University. She is the author of more than 150 papers and more than 13 patents and has presented over 200 technical papers at conferences. She has served on advisory boards at the National Academy of Sciences, National Science Foundation, the Department of Defense, and several professional societies and was a member of the President's Committee on the National Medal of Science. She enjoys skiing, diving, backpacking, reading and singing choral music, interests that have brought her to interesting places such as: Belize, Bora Bora, Turks and Caicos, Thailand, St. Peter's Basilica in Rome, and Carnegie Hall.

The Early Years

Growing up, I was a word omnivore. If it was in book or

9

magazine form, I read it, which made me a sometimes obnoxious know-it-all brat. My father was the first person in his family to graduate from college. His degree was in accounting, but he was an avid science fiction fan. Also, as an Army officer, he needed to understand technology, so he subscribed to *Popular Science* for many years (of course I read it, too). As the oldest child, I spent time with my dad learning how to use tools and was often "volunteered" to help him in his shop.

I remember going to the post library in Panama when I was in high school and coming home with sci-fi books for Dad, novels for Mom, children's books for my little brother and sister, and miscellaneous fiction and nonfiction books for me. I would read them all before returning them to the library.

In high school, I had an intrinsically interesting math teacher: Mrs. Graham. Mrs. Graham had a master's degree in math from Hunter College in NYC. Every six weeks, she would assign a report on "anything that had math in it." I remember writing papers on nuclear physics, probability and statistics, nonlinear geometry and celestial navigation, using my dad as a sounding board when I got stuck. Mrs. Graham did not put up with any nonsense, had high standards, and was obviously respected by her students and peers. I learned a lot from her and from writing those papers. Mainly, I learned that I was capable of tackling hard problems.

While I was interested in science and math in high school, it wasn't my favorite subject, as I loved books (and still do). I remember taking an aptitude/interest test and being given the choice between "baking a cake" and "taking apart a clock." I picked the "taking apart clocks" type of answer every time. The test results said I should aspire to being a medical technologist. I was outraged and maintain to this day that, if I had been male, the test would have predicted that I should be a medical doctor. It all worked out for the best, as a sense of righteousness and a stubborn streak contributed to my

10

deciding to major in science in college.

Higher Education

I went off to college at Duke: the university that I still consider my alma mater. I decided to major in chemistry (rather than English or engineering) for two reasons: first, there were too many English majors, and I didn't like dissecting my beloved books; and second, I wasn't brave enough to be one of just a scant handful of female engineering students. I had no plans for what I would do after college. The best I could think of was to use my parent's contacts in the military to travel by finding a secretarial job overseas for a few years.

I met my husband, Chuck, on a blind date arranged by my mom my first year at Duke. Chuck and a friend violated standard military policy and volunteered to serve as dates for my roommate and me. We fell in love and were engaged later that summer. Just prior to our engagement, Chuck finished his military school and was transferred to Ft. Carson, CO. It was much less feasible to have a long distance romance in those days. Air travel was relatively more expensive, and the distance between North Carolina and Colorado seemed very large. We were married during Christmas break, and I went back to Duke after our honeymoon to take finals – fun! I then transferred to Colorado College for second semester of my sophomore year. I finished my undergraduate studies with a summer session at UC Berkeley (the only place I could find a required chemistry course in summer session) after just 3 years, living in the dorm while Chuck was on maneuvers most of the time. Burned out, I decided to get a master's degree at Cal State LA and teach at a junior college. I discovered that I loved graduate school – the opportunity to really concentrate on one thing and apply what I had learned in class in the research lab. Not wanting to spend 3 years getting a master's degree in chemistry and having aced my second attempt at the GRE, I applied to local Ph.D. programs. I didn't get into UCLA because of my GPA (brought down by an F received after coming back from our honeymoon), but I was accepted into USC and Cal Tech. I chose USC because they offered me an immediate assistantship, whereas Cal Tech's schedule started one month later and they would not make decisions on assistantships until after the deadline to reply to USC.

While interviewing potential research directors, Otto Schnepp, who would become my advisor, asked me if I knew what a Phillips head screwdriver was. Turns out all those hours helping my dad in his workshop actually paid off!

I was a few months pregnant when I took my qualifying exam. By design, none of my professors knew, as pregnant graduate students were not necessarily looked upon kindly in those days. Our son Hal was born in 1969 while I was finishing the synthesis of the molecule, the spectrum of which would be the basis of my dissertation. My main research was the design and construction of the vacuum ultraviolet circular dichroism spectrometer that took the spectra. I had a lot of help with the synthesis! In addition to coursework, I learned a lot of practical things in graduate school: how to design parts to be made in the machine shop, glassblowing, welding, vacuum and electronic equipment, and how to work 36 hours straight. The most important thing I learned was how to tackle almost any problem; being a trained problem solver is definitely handy.

Chuck was, at this time, going to law school at night while working in management at Sears. It was tough not seeing each other, but we made it. Chuck finished law school the same month I received my Ph.D. I give Chuck, an attorney by training, a lot of credit for my academic success. He has always remained supportive and continued to push me to achieve because, as he said, "If you understand all that math and science stuff, you must be pretty smart."

Early Career

We set out to find two jobs. Chuck wanted to live in Colorado (where he had earned his undergraduate degree) or Iowa (where his family was from and where he grew up). Unfortunately, I couldn't get a job in either state. In the early 1970s the academic world was not a very viable option for a woman scientist. I wasn't willing to accept a teaching-only position, and research faculty positions for women in university chemistry departments were limited. After trying to find jobs simultaneously, we decided that I would find a job, and then Chuck would take the bar in that state and find a job. I received job offers from Western Electric in Princeton, NJ, and Hughes Research Labs (HRL) in Malibu. After much discussion, we decided to stay in Los Angeles, and I took the HRL offer.

HRL was where I became a laser jock. Although I had worked with lasers as a grad student and during my short time as a postdoc at USC, the group I went to work with at HRL was looking at high power laser damage to optical components. I learned a lot and had a good time, but after about four years, the atmosphere in the group I worked in changed, and I decided to look for another job. I ended up back at USC as a research scientist in the Center for Laser Studies – magically transformed overnight into an electrical engineer. With help from Elsa Garmire, another essayist in this book, also at USC at that time, I learned how to write proposals, received my first grant, and established my own research group. I was reasonably successful, but paying my salary was taking a big chunk out of my grants, so I decided to look for a tenured position. The first time I did so, I received mixed responses: chemistry departments were fine with my background but didn't know what to make of my research. Engineering departments were happy with my research but always wanted to know something like: "Can you teach circuits?"

By 1987, the climate for interdisciplinary work was more favorable. I received 5 job offers: an endowed chair in electrical engineering and full professorships in electrical engineering, materials science, physics, and chemistry. The chemistry

13

position was at the University of Iowa, where they were starting a new program in laser science and engineering. Seizing the opportunity for Chuck to return to his home state, I took the chemistry position. I never did end up doing the standard tenure thing – my first tenure track position was as a tenured full professor.

During this time our daughter Eleanor was born. With two incomes, we could afford to add an apartment onto our house and hire a live-in nanny. Most of the time this arrangement worked well, and Chuck agreed that he would take on the chores for Eleanor, as I had been responsible for most of the childcare duties when Hal was young. Eleanor still talks about how she was the only kid whose dad made her lunch and took her to appointments!

The Price of Popularity

I was a very high-profile hire as the first faculty member in the Center for Laser Science and Engineering and the first female faculty member in chemistry since the founding of the University of Iowa in 1847. I was asked to do a lot of public speaking – everything from Rotary to dinner speaker to half-time at the basketball game, and I learned I could do this without fainting from nervousness. I did ask, after a while, that they stop introducing me as the first-ever female faculty member in chemistry. I didn't really think it was something they should be proud of.

Because of my name recognition, I got elected to everything! I served on the College of Liberal Arts Executive Committee, where I discovered that university administration, contrary to popular belief, really could effect change and that a faculty member could, if in the right place at the right time, actually help. I was also elected and appointed to the Board in Control of Athletics (BICOA), the Faculty Senate, the Senate Executive Committee, and numerous university committees.

I was an advocate for women faculty, particularly in underrepresented Science, Technology, Engineering, and Mathematics (STEM) fields. I, along with some other science and math faculty members, organized a science faculty group called NEW – Network of Experienced Women – comprising less than 12 people when we started and growing to more than 20 when I left five years later.

My research group was doing well, and I had a lot of grant money. We were working on several projects, including looking at the mechanisms of laser chemical vapor deposition (LCVD) and laser particle removal (LPR). This research helped address a major problem for semiconductor manufacture and other critical surfaces such as high power lasers and telescope mirrors, especially in space. Feeling a need for new challenges and with the encouragement of Dean Gerry Loewenberg, I decided to start applying for administrative positions.

Administrative Experience

The problem with having one foot in science and one in engineering is that you are not eligible for dean positions in either. I eventually ended up with a position as dean of the graduate school at Tulane. I was also the Vice President of Research for the main campus, charged with organizing a new joint research program with Xavier University, a private Historically Black University (HBU) in New Orleans.

After four years at Tulane, I accepted the position of Vice President for Research at Florida State University (FSU) in 1996. This was a much bigger job with a significant budget. It was a very steep learning curve with new pots of money and new skeletons being discovered every week for the first 6 months. I did manage to get the budget under control and hire several good staff but did not "connect" with the president at that time, so I stepped down and went back to the faculty.

Is there life after administration? Definitely. My lab equipment was still in storage at Tulane so we had to get it to FSU, rehab some space, unpack, and start hunting money. I was lucky enough to learn about a Defense Advanced Research Projects Agency (DARPA) program and received funding within the first year after returning to the faculty. I built my program from there.

Eventually, I returned to administration. Arkansas State University (ASU) had been looking for a Vice Chancellor for Academic Affairs for almost three years. The search committee authorized a hunt for someone with research administration experience, to help start a major research effort at ASU (traditionally a teaching institution). A headhunter called me and I said, "Where is Arkansas State? I've never heard of it." It turns out that Les Wyatt, the president of the ASU system and the then chancellor of the Jonesboro main campus, is very persuasive! He convinced me that there was going to be significant growth at ASU and that I could help. Together we increased research funding at ASU almost an order of magnitude and established the research infrastructure to support faculty. It was a wild ride. I learned a lot about campus and state politics and enjoyed working with Congress and starting the ASU branch of a statewide bioscience research initiative. At the same time, Chuck and I celebrated our 40th wedding anniversary and moved into a 100-year-old house. I kept a lab and hired a colleague to help. Our group of five laser-jock faculty worked well, and I got to "do science" occasionally.

A new chancellor was hired, and I again went back to the faculty while helping the new chancellor during his transition period. We had several million dollars of research funding for the "Laser Jock" group (seriously, that was the name listed on the ASU alphabetical listing), so going back to the lab full-time was a bit of a relief. I really was working more than two jobs for a while. As Distinguished Professor and Director of the Arkansas Center for Laser Applications and Science (ArCLAS), I worked on development of laser-based sensors for explosives. I also started a biology project using ultraviolet light emitting diodes (LEDs) to kill bacteria with and without a wide bandgap semiconductor to provide photocatalysis. All of these projects began from scratch and are just beginning to produce peer-reviewed publications at a regular rate. An active inventor, I was awarded three patents in the last several years and have new ideas in the works all the time.

With another change in administration, there was a change in the climate for research at ASU, and I started a low key search for a new job, particularly in my home state of Florida. As support for research at ASU deteriorated and major research programs were in the process of being dismantled, the position of Distinguished Professor of Mechanical Engineering and Associate Dean for Research at Embry Riddle Aeronautical University (ERAU) in Daytona Beach, FL, was very attractive. ERAU wants to grow its research and graduate programs, and I like helping to do that. They are also interested in recruiting more women and minority engineering students to this unique university, and I hope to help with that mission. As of this writing, I have been at ERAU for a little more than a month. It is a very dynamic institution with highly committed students and faculty, and I think I will enjoy being here.

Reflections

Looking back at my ambitions in high school and college, things certainly didn't turn out the way I expected them to. Majoring in engineering was not an option, but now I am a professor of engineering. Professorships were not an option when I received my Ph.D. – there were many chemistry departments with no female faculty at that time. I never knew about, never mind considered, being chief academic and research officer for a university. But times have changed, and the options for women have expanded. Things still are not always easy, but the options are there.

I definitely do not recommend my random walk approach. I

think that young women and men should plan their careers and should aspire to great things. That said, always keep your mind open to new possibilities in your career, just as you keep your mind open to new ideas in the lab and classroom.

My proudest accomplishments? Our children – Hal is, like his father, an attorney. He trained as a barrister in Great Britain and has a LLM from Columbia. After working on Wall Street in NYC, he now trades the market himself. Eleanor graduated from Barnard with a degree in biology and a chemistry minor, continuing her education with a Ph.D. in pharmacology from Weill Cornell Medical College in 2012. She began a postdoc in chemical biology and tuberculosis (TB) drug discovery at SUNY Stony Brook in summer 2012. I am also especially proud of my marriage and, of course, my students – postdocs, graduate, and undergraduate. Each of my students and postdocs are special in their own way, but I am particularly proud of the undergraduate students I have worked with who decided at some point that mediocre grades were not sufficient and went on to academic and professional success.

And – after a long random walk, I'm still standing!

ANI APRAHAMIAN

Bona Fides

Dr. Ani Aprahamian is a Frank M Freimann Professor of Physics at the University of Notre Dame. She is the co-chair of the Science Academies Decadal Review of Nuclear Science. She is a member of the Science Academy of the Republic of Armenia and is a fellow of the American Association for the Advancement of Science (AAAS) and the American Physical Society (APS). Ani has served on many national and international committees advising, reviewing, and planning for small and large science projects alike. She is a member of the Joint Institute of Nuclear Astrophysics (JINA), a physics frontier center funded by the National Science Foundation (NSF). She has served at the NSF as a program officer in Particle and Nuclear Astrophysics and Nuclear Physics. Ani is presently the chair-elect of the Division of Nuclear Physics.

Nuclear Scientist and World Citizen?

The Old World

It was July 1969; the Peter and Elizabeth Torosian elementary school located in one of the suburbs of the city of Beirut had just closed for the summer. My maternal grandmother was taking me along with her to the bazaar that morning. I was elated...the Americans had landed on the moon! While we did not own a TV set due to the strict instructions of our fundamentalist school, the newspapers were covered with photos of Armstrong walking on the moon. My grandmother, a very religious woman, was agitated and reciting prayers asking God's mercy on earth to forgive the people who had dared to interfere with the heavens. That morning walk with my grandmother and our divergent views on the events of the day is the perfect metaphor of living in Lebanon in 1969 and the foundations of my deep interest in science.

Our family had come to Lebanon following the first genocide of the 20th century, when the Turks tried to eliminate all Armenians within the current borders. My brothers, sister, and I were amongst the lucky who had two sets of grandparents. Our grandparents had walked through the Syrian desert of Der El Zor and lived in a refugee camp on the beaches of Beirut. The camp consisted of tin and wood one- or two-room cottages with muddy streets and no running water. My parents were educated through 6th grade in schools founded by missionaries from the U.S. and France, and they were both apprenticed to master tailors. My father was apprenticed to a respected men's tailoring house in downtown Beirut and my mother to a women's tailoring mistress in the neighborhood. They used their skills to get out of the "camp" and move into stone buildings in one of Beirut's suburbs.

Lebanon was then – as it is now – overwhelmed with religious and political views coming from the traditions of Christianity, Islam, and Judaism. The elementary school I attended included close study of the Bible. At 10 yrs old I had studied all parts of the Bible including the book of Revelations, which specifically predicts Lebanon becoming "a stable for the horses of Israel." I studied this just two years after the defeat of Lebanon in the 1967 War with Israel. Family discussions at breakfast, lunch, dinner and the cozy extended family gatherings in the evenings dwelled on one topic alone: the future of Lebanon and whether this defeat was just the realization of predictions made a long time ago. There were uncles who argued the opposite point, that the Bible had nothing to do with current events. My grandparents were not so sure. Some cousins cited Marx, others preached capitalism, and others were stuck to religion.

My head was full of questions: of origins, history, religion, and science:

Was religion really the opiate of the masses?

How did we get here?

Who are we? Lebanese? Armenians? Phoenicians? Anatolian tribes?

Is there a different explanation for everything?

What do mathematics, physics, and chemistry say about how we got here?

How do things work?

How does water boil? How do snowflakes form?

How do the stars shine at night?

Was there really a "pre-destiny" like the ancient Greeks believed?

Was our world the evolution of many former worlds?

The political and religious turmoil in the country was coupled to learning at our elementary school, which was built by two American benefactors. Elementary school education in Lebanon included introductions to biology, chemistry, geography, and physics as well as religion, history, and literature taught in three languages (Armenian, English, and Arabic). Whenever possible, I tried to determine the facts and really understand what was happening all around me. Needless to say, it was a naïve approach to the complexities surrounding us, but somehow this fact-based approach provided me with inner strength, strong convictions, and determination to survive and to help others do the same. This period is when I

21

became convinced that science would provide the answers to life's questions free of bias, superstition, and religion, a founding principle for my life.

The benefactors had provided a library full of many illustrated science books about the lives of plants, animals, the universe, and the elements. I spent many of my free hours in the library absolutely fascinated with these books, and also awakening my lifelong love of art and everything visual and aesthetically pleasing.

Two years later, my family immigrated to the USA. We moved to the city of Worcester in the state of Massachusetts. We left behind a large extended family of more than 40 first cousins, many aunts and uncles, and grandparents. My parents were often sad and lonely having left behind the entire clan to provide us a better chance at education. My father had told us on the day we left for the USA, that all five of us would now be able to graduate from high school. My mother and father had a thirst for learning that they were not able to quench. Their only continued education was to be through newspapers and magazines in English and Armenian.

I have often felt guilty about not feeling sad to leave the family behind, but I had an indescribable happiness that we were going to the country that did science! The moon landing had made its imprint, and all the science books in the library provided by our American school benefactors convinced me that the U.S. was a country of science, technology, and the future.

The New World: Coming to America

My path has always been paved with incredible teachers, friends, and acquaintances. The summer of our arrival in the USA, I was signed up for some English-as-a-second language classes. The classroom was filled with Greeks, Italians, and Arabs; all of us were new immigrants. Our teacher was Mr.

Carpenter, who appeared to represent all that is good with Americans. He was dedicated, caring, open-minded, and devoted to evaluating our correct placement levels. He decided that given the math and science that I knew, I should be promoted to 7th grade instead of 6th. He was the first person to help me in my American educational journey. In junior high, I learned French. Equipped with the ability to speak 5 languages, I dreamed of being a career diplomat – helping the world to solve conflicts without going to war. My junior high guidance counselor pointed out that having no political connections to any past, present, or future U.S. presidents, my chances were not so good for becoming an ambassador. She instead encouraged me to pursue science and mathematics, my best subjects.

I was then enrolled in the best-reputed high school (Doherty Memorial) in Worcester, with a selected group of Honors students. My classmates were impressive, and I learned so much from them. They told me about taking standardized examinations (SATs) and applying to colleges. My classmates applied to Harvard, MIT, Caltech, and Stanford. My own goals were modest; I intended to be trained in something that would allow me to have a job and support my family. I had somehow chosen to be a dental hygienist and asked my physics teacher, Mr. McCann, for a letter of recommendation. Mr. McCann asked me some questions about my future plans and said, "Over my dead body, Aprahamian, are you going to a dental hygiene school!" On his suggestion, I applied to Clark University in Worcester and got in with a full scholarship. I majored in chemistry, my first love, and was fortunate in my junior year to have Prof. Daeg Brenner as my instructor for Physical Chemistry I and II. Prof. Brenner was faculty in both physics and chemistry at Clark University, with a specialization in nuclear science. I realized that I loved the physics basis of chemistry and studied nuclear science the following year with great enthusiasm. Upon graduation, I completed a summer research project with Prof. Brenner, working to build the excited structure of ^{183}W. I loved the work and considered going to graduate school, but all deadlines had passed! I mentioned my interest in graduate work, and Prof. Brenner whipped out an application and said "sign here!"

Daeg Brenner took me to Brookhaven National Laboratory on Long Island (NY) to do research over a summer, opening up the world of nuclear physics to me. At Brookhaven, Richard Casten became my mentor and daily advisor. We studied the fission of ^{235}U at the High Flux Beam Reactor at Brookhaven. Dr. Casten would give me daily lectures in nuclear physics. He eventually wrote a textbook based on his lectures to me called "Nuclear Structure from a Simple Perspective," presently in its second edition. At Brookhaven, I had the opportunity to meet and work with incredible scientists from all over the world in addition to Rick Casten and Daeg Brenner: David Warner, Kris Heyde, Franco Iachello, Roelof Bijker, Alejandro Frank, Stuart Pittel, Birgir Fogelberg, Peter Moeller, and many others. At BNL, I assisted on a reactor experiment for Dr. Richard Meyer of Lawrence Livermore National Laboratory and was offered a job as a postdoctoral fellow to Livermore.

I graduated with my Ph.D. in the first week of August 1985, gave my first invited talk at an ACS conference in St. Louis the following week, and then flew to San Francisco to start my postdoctoral fellowship.

At Livermore, I learned about data acquisition systems from John Becker, about detectors from Wolfgang Stoeffl, about analysis from Gene Henry, about culture from Lloyd Mann, about new physics from Hugo Maier, and received useful advice from Daniel Decman and Pat Mann. All the training was invaluable to my present job at the Nuclear Science Laboratory at the University of Notre Dame.

Dr. Franco Iachello of Yale University had been invited in 1988 to give a colloquium talk at Notre Dame. During his visit, he mentioned to them that I might be available. I was invited by Prof. Umesh Garg to come to Notre Dame to give a talk, and two weeks later I was offered a job as an assistant professor.

Now?

I feel very fortunate for all the likely and unlikely events that strung together to yield my professional path. Even though I had already worked at a number of national laboratories in the USA and overseas, I had very little experience running accelerators (formerly I had access to professional operators). As an assistant professor at Notre Dame, I learned to operate the accelerator alongside my very first two graduate students and had a lot of help to launch my career, notably the invaluable advice and mentoring of James J. Kolata and Michael Wiescher.

Now, as an endowed chair professor of physics at Notre Dame, I still hold my original conviction that science and our science colleagues across the world hold the key to peace. In 1989, after glasnost, my LLNL colleagues knew their counterparts in the former soviet republics and were most concerned about their welfare. I believe that our passion for a common interest in science in general and nuclear science in particular can help us overcome many of our national, racial, and ethnic borders. Working with colleagues from over 50 countries to date, I think the scientist is the true world citizen: "the Cosmopolitan."

Advice for the Next Generation

My path has not been easy, but it has been enormously rewarding. Compared to my male colleagues, I sometimes had a harder time getting new ideas accepted, and some things took much longer, but the real reward was gaining an understanding of a little piece of the universe that was previously unknown. Science was and continues to be a lifeline for me from the tumultuous Middle East to the 21st century. A very good friend of mine once gave me a poster with a bicycle on it that read, "Follow your heart wherever it may lead." I thought it frivolous then but now believe in it more than ever. Find your passion and follow it. Look kindly on those who criticize you and support you. After my 23 years as a tenured professor, I can see that nothing I did was a wasted effort. I learned from failures as much if not more than my successes. I wish to keep learning and contributing to our universal consciousness in my own way, and I wish it for you as well.

SHEILA BROWN BAILEY

Bona Fides

Dr. Sheila Brown Bailey graduated from Duke University with a Bachelor of Science degree in Physics and then completed her M.S. at the University of North Carolina at Chapel Hill. Sheila completed her Ph.D. in Physics at the University of Manchester in England and subsequently traveled to Australia to complete a postdoctoral fellowship at the University of New South Wales Royal Military College. She is currently a senior physicist doing photovoltaic research at NASA Glenn Research Center. She has been the recipient of the NASA Exceptional Service Medal for Space Photovoltaics and numerous NASA Special Act or Service Awards. She has completed OPM's Executive Potential Program and is an Ohio Academy of Science Exemplar. In 2003, she was inducted into the Ohio Women's Hall of Fame by Governor Taft. She has served for over twenty years as the Executive Vice President of the Lewis Engineers and Scientist Association, IFPTE Local 28. She is a strong participant of the NASA Glenn Speakers Bureau and the American Physical Society Visiting Scientist program, focusing both on space photovoltaics as well as careers in engineering and science.

The Early Years

My mother met my father during World War 2. He was in the Navy, and she was finishing her RN at the Virginia Commonwealth School of Nursing in Richmond, where I was born. After the war, they returned to my father's hometown of Landis, North Carolina, where I grew up. I enjoyed music and sports in high school and was voted "Most Athletic" in the Yearbook after becoming an All-Star basketball player and captain of the tennis team. I always loved math, physics and chemistry and had a wonderful high school science teacher, Patricia Barrow, who encouraged my participation in science fairs and summer science programs. My high school project on radioactive fallout from the Russian atmospheric nuclear tests won a prize from NASA at the N.C. State Science fair. After completing a summer NSF program in astrophysics at Emory and Henry College, I was determined to major in physics in college.

Even though my father only had a high school education he was very determined to see me attend college. Getting accepted to Duke University with only a limited rural southern school education was a challenge – there were no advanced placement or calculus classes offered at my high school. I struggled the first year in college but discovered that applying a strong work ethic could help me catch up with my better-educated classmates.

I married my sophomore chemistry laboratory instructor the day after my college graduation. I attended the University of North Carolina (UNC) at Chapel Hill to earn a Master's degree while my husband finished his Ph.D. at Duke. At UNC, I published my first two research papers under the guidance of my very supportive Masters supervisor, Dr. E. N. Mitchell. The first paper was on a practical way to measure low temperatures by evaporated silver-aluminum thermocouples, and the second was on unidirectional anisotropy in permalloy films at 4.2 degrees Kelvin. The latter research was related to the early problems of magnetic memory for computers.

My husband and I managed to finish our degrees at the same time, and so we headed to Europe to attend an international summer institute in quantum chemistry and solid-state physics in Norway and Sweden, supported by a NATO fellowship. He started a postdoctoral fellowship in the University of Manchester's chemistry department as I completed my Ph.D. in physics under the supervision of Dr. M.A.H. McCausland. Dr. McCausland was a terrific mentor. My Ph.D. thesis was titled "Environmental effects on the hyperfine interactions of rare earth ions in magnetically ordered systems," which involved hyperfine and relaxation measurements taken at $1.4°K$ of the probe nucleus holmium in rare earth alloys and intermetallic compounds. Thanks to a liberal policy of 6-week vacations, we were able to travel extensively in Europe and North Africa using just a car and tent.

Unfortunately, my husband and I didn't quite manage to finish our programs at the same time – my husband waited about six months after his postdoc for me to complete my degree. Loving travel, we decided to move to Australia for postdoctoral fellowships at the Royal Military College. My sponsor was Dr. G.V.H. Wilson, the head of the physics department. The college, affiliated with the University of New South Wales, offered undergraduate degrees to the Army cadets who trained there – the graduate degree program was just instituted as I arrived. It was rather strange to be the only female faculty member and therefore the only woman member of the officer's mess. However, both the students and faculty were welcoming. I was even taught to curtsey properly in order to meet the Governor General of Australia, Sir Paul Hasluck. My research focused on hyperfine fields and line broadening mechanisms in dilute cobalt in iron alloys and scandium hyperfine fields in various hosts. I was extremely fortunate to have both accomplished and supportive supervisors throughout my academic training.

Around the World and Back Again

We would have stayed another year in Canberra, but my husband was unable to find suitable employment. So, we spent the next year and a half traveling. First we hitchhiked to

Darwin then flew to East Timor and journeyed overland across Bali, Java, and Sumatra; we then hopped on a boat transporting raw rubber to Singapore. After that we traveled through Malaysia to Thailand. Finally, leaving Bangkok, we flew back to Manchester. Before leaving England we had purchased an old British ambulance and left it with friends in Liverpool. We returned to stay with them a few months, converted the ambulance into a camper, and then drove from Liverpool, England to Cape Town, South Africa (taking a few ferries along the way). We planned to ship the van to India from Durban, SA and then drive back to England, but that plan was revised when I discovered I was pregnant. I flew back to the U.S. from Johannesburg seven months pregnant. Our son was born in Salisbury, N.C. in July of 1974.

We both found jobs teaching at the University of North Carolina in Charlotte for a year and then moved to Delaware, where my husband taught at Swarthmore College, and I was at the University of Delaware. My husband then took a job with Eveready Batteries, and we moved to the Cleveland area and lived on a small farm.

Our first daughter was born in 1977 followed by our last daughter in 1978. During this time I taught part-time at Cuyahoga Community College and Lorain Community College. In 1979, I took a sabbatical replacement offer at Oberlin College. I don't think I was a particularly good teacher in the beginning, but I enjoyed it, and I listened closely to student feedback. Most of the courses I taught were designed for non-major students, so I had fun teaching courses like "How Things Work," "Light, Color, Vision," "Einstein," etc. In 1981, I began teaching at Baldwin Wallace College and remained an adjunct faculty member in their weekend college for 27 years, teaching courses on "Physics in the Everyday World," "Space from the Ground Up" and "Renewable Energy." I was very proud when I won the Faculty Excellence Award in 1988, since many of the votes came from business students who were taking my disguised physics courses as a laboratory science requirement.

My Career at NASA

In 1984, when my children were 5, 7, and 9, my life fell apart after my husband left me for a woman from work (whom he'd later marry). In reviewing my options for full time employment I inquired about positions at NASA Lewis Research Center. I was particularly interested in renewable energy and interviewed with a new supervisor, Dr. Dennis Flood, in the photovoltaic group at NASA.

Government jobs are often a matter of timing. I was fortunate to be offered a position in that group, particularly since I had no background or experience in the area. But physicists are nothing if not versatile! I was given the opportunity to take classes at Cleveland State and Case Western Reserve University to bring my skills in semiconductor physics up to date. Dr. Flood, a wonderful mentor, had confidence in my ability to design laboratories and experiments that would contribute to the broad goal of making a "better" space solar cell. He was very supportive, advocating for my inclusion in educational opportunities like the International Space University and for my serving on the organizing committee of the Photovoltaic Specialists Conference. To date, I remain the only female chair of that conference, serving as chair of the 4th World Conference in Photovoltaic Energy Conversion in 2006.

The worldwide photovoltaic community, while predominately male, contains some remarkable women. It has been a pleasure to form friendships, both male and female, around the globe. I'm also grateful that working at NASA has provided many more chances to travel, including serving as faculty at the International Space University in Thailand and Chile and participating in conferences in Japan, Korea, China, and Europe.

It is exciting to participate in experiments that actually land on

Mars or fly to the International Space Station. I often ask students: Where else are you paid to play with very expensive toys (such as my field emission scanning electron microscope)? I am absolutely positive I would hate a repetitious, routine job. I enjoy the challenge and excitement of charting my own course.

I have also enjoyed my 20+ years as the lead for the Scientist and Engineers Pact of the International Federation of Professional and Technical Engineers Local 28 – work that I consider community service. I am treasurer of the NASA Council of Locals and participate on the Agency Labor Management forum as well as the Center Labor Management forum. One of the most important aspects of the job is our right to lobby on behalf of our workforce. Without the right to bargain for salaries or benefits or to strike, the union is primarily concerned about the working environment. I have always felt that collaborative efforts up-front yield more positive results than confrontational approaches after the fact, so we maintain close legislative contacts to ensure that our Center's best interests are protected. Many scientists often regard politics as a completely baffling entity, but it has taught me that short concise sentences translate best. This has carried over into my technical work as well. I laughingly tell students that the reason NASA always has so many pictures is that we spend a lot of time talking to upper management and Congress – sort of like "show and tell" at school.

Choose Your Future Wisely

Entering a research world after being absent for more than ten years was challenging for me. However, I was fortunate to have wonderful colleagues in the Photovoltaic group and an excellent boss. Most importantly, I must give credit to my second husband for the very supportive role he played, particularly during my children's teenage years.

I rarely found discrimination among my peers. Occasionally I would encounter condescending attitudes from some men – usually older – at NASA. They would look like they wanted to pat me on the head and say "Poor dear, she can't be expected to understand." Some actually tried it. I didn't take that well as a young woman but found that as I aged my response became more focused on educating the men. Interestingly, the only real impediment I encountered in my career stemmed from a female boss after Dr. Flood retired. Some women, just like some men, are incredibly ill-suited for management.

I have never been sure if my direct approach to discussions or problems emanated from being in a primarily male environment most of my life or from my own personality. It seems that the younger women at work are sometimes hypersensitive to criticism and often take offense where none was intended. No one can escape the social components of a workplace – differing backgrounds, religions, politics, education, mannerisms, defensive and offensive postures, etc. I've found that sometimes in meetings it is helpful to establish "rules of order" – at least then everyone knows and can play by the same rules.

The path to a career is not always straightforward. Take my children, for example. My son received his Bachelor's in physics from Stanford and now works as an investor in New York City. My oldest daughter received a degree in biomedical engineering and is now a medical doctor. My youngest daughter double majored in environmental science and English and subsequently obtained an M.A. in International Environmental Policy and another M.A. in Regional Security Studies. She is still looking for her way in life.

Several years ago I was asked to talk to a young woman who just finished her first year in college. She was trying to decide on a major. She had an interest in physics from high school but had not taken any college physics courses. She spent a day with six Ph.D. physicists at NASA GRC and myself. There were a huge variety of activities that we seven supported, from aeropropulsion to sensors for the international space station. I was pleased to discover that she graduated this year summa cum laude with a double major in physics and English. For those of you who are reading this and trying to decide what to do with your life, remember that having options is always a good thing. A degree in physics provides a wonderful array of choices for the future.

To those of you pondering a career in science, math, or engineering – ask yourself these following questions:

Are you curious? Do you ask questions? Do you seek answers or just wait for someone to answer you?

Do you organize your time, both work and play, or do you wait for someone else to tell you what to do?

Do you take that extra step? Do you do that science project or extra-credit task?

Do you take those harder courses, knowing that it will be more work and perhaps not as good a grade?

Do you prefer new experiences rather than a routine?

Do you accept responsibility for your mistakes or blame others? Do you learn from your mistakes?

If so, you'd make a great scientist/engineer/mathematician. Here are some of the next steps:

Take that algebra/geometry/calculus course (even if it isn't fun or easy).

Take **ALL** those science courses.

Don't give up if you don't get an A.

Believe in yourself. You CAN do it if you keep trying.

Don't waste your summers: NASA, NSF, NIST, and DOE all have summer programs for high school students and undergraduates. Think ahead since many of these programs have deadlines for summer applications in *January*.

Maximize your potential to have the job that YOU want. You will have been in school 17 years by the time you graduate from college. You will work on the job for an average of 45 years after that.

There are two quotes that I most often use in discussing careers with students. The first is by Robert H. Goddard: "It is difficult to say what is impossible, for the dream of yesterday is the hope of today and the reality of tomorrow." Want proof? Here are some very famous people making some very dubious claims:

1895 - Lord Kelvin: "Heavier than air flying machines are impossible."

1897 - Lord Kelvin: "Radio has no future."

1901 - Wilbur Wright: "Man will not fly for fifty years."

1932 - Albert Einstein: "There is not the slightest indication that nuclear energy will ever be obtainable. It would mean that the atom would have to be shattered at will."

1943 - Thomas Watson, Chairman of the Board of IBM: "I think there is a world market for about five computers."

1957 - Dr. Lee DeForest, Inventor of the Audion Tube: "Man will never reach the moon regardless of all future scientific advances."

The second quote I often use addresses what it takes to make a difference in the world around us: "Few will have the greatness to bend history itself, but each of us can work to change a small portion of events, and in the total of all these acts will be written the history of this generation." This was said by Robert F. Kennedy.

I am happy that my research will benefit our future world, and I have thoroughly enjoyed doing it. I wish you all the same in your careers.

OLGICA BAKAJIN

Bona Fides

Dr. Olgica Bakajin is a Chief Executive Officer at Porifera, Inc., a San Francisco Bay area startup company dedicated to the development of advanced membranes and systems for water treatment. She received her Bachelor's degree in Physics, with satisfied requirements for a Bachelor's degree in Chemistry, from the University of Chicago in 1996. In 2000 she received a Ph.D. in Physics from Princeton University. She then joined Lawrence Livermore National Laboratory (LLNL) as a Lawrence Fellow. After spending the first 9 months of her fellowship visiting the National Institutes of Health, Olgica came to LLNL. She conducted independent research there, until 2003 as a Fellow, and from then until July of 2009 as a member of the permanent scientific staff. Olgica led her own research group at LLNL, which discovered fast water and gas flow through carbon nanotubes. She has co-authored 6 issued US Patents and several patents pending, and her 40 peer reviewed publications have been cited over 5000 times. She has been a PI on many government-funded projects amounting to more than $14M and was the only female PI among the inaugural ARPA-E awards. Her other honors include the NanoTech Briefs Award in 2007, the R&D100 Award in 2010, and election as a Fellow of the American Physical Society in 2011.

Growing up in Yugoslavia

Growing up in a socialist country gave me a boost towards a

career in science, but possibly steered me away from other possibilities. I was always good at math and logical reasoning, but my memory for facts was always bad. I liked physics because I could always derive everything from the first principles. The sciences, unlike the humanities, weren't political, so they were taught well in Yugoslavia. History was a skewed pile of dates and names, so I never became a historian. Economics was stale Marxism, so I never considered becoming an economist. Math and science seemed much more interesting. From 5th grade throughout high school I participated in and won prizes at math and physics contests. I went to various geek camps for talented kids that the state was sponsoring. All of the smartest kids were channeled into studying math and science, and that's where I found like-minded friends. Recognizing my own skills early on was an important boost to my own confidence. Connecting with other nerdy kids kept me from being an isolated outsider. Being in a socialist country meant that even girls were encouraged to do math. Luckily for me, I also liked to write. In high school I was a regular contributor to the kids section of the city newspaper. I learned how to craft a story, which turned out to be a really important skill.

Being smart and having interests outside science has been important, but having a supportive family has been a critical part of my success from the beginning. My parents have always been supportive, and they created opportunities for me. For example, they sent me to England to learn English, which was a financial sacrifice for them. My parents treated me as a responsible individual from the very start, and I always felt like my opinion mattered in what our family did. I planned my own days; I roamed on my bike around the city; I was allowed to be independent. I was able to develop self-confidence not only in my academic pursuits but in my daily life as well. Nowadays my friends remark on my confidence. Is my lack of insecurity due to something my parents did, or was it going to be there anyway? Now, as a mother, I wonder about ways to create that atmosphere for my own kids.

My life changed when I was chosen as one of the two representatives of Yugoslavia at the Research Science Institute, a science program organized by Center for Excellence in Education. I came to the U.S. for six weeks and hung out with math and science geeks, mostly from the U.S. but also from other countries. I loved the people. I loved the program. I loved that I was in the U.S. It was the best time of my life. Crucially, at RSI I saw the next important step in my life. While at RSI I learned that U.S. colleges give financial aid to foreigners. Although we lived comfortably in Yugoslavia, my parents would never have been able to afford to send me abroad to college, so I had not seriously considered it. Once I found out about financial aid, that was it: I was determined to get into a college in the U.S. The gathering signs of war in Yugoslavia and the nationalism ripping the country apart made it that much more urgent. I went home and studied for the SATs and TOEFL and did whatever I could to increase my chances. I was lucky enough to get three acceptance letters. Both the University of Chicago (U. of C.) and Brandeis offered a full financial aid package, whereas MIT offered a student loan. With my parents blessing, I picked the U. of C., left my family, and went to Chicago.

College and Grad School in the U.S.

At U. of C. I learned a lot both inside and outside the classroom. I studied hard but found time to go out to jazz jam sessions in clubs around town. I also loved going to Lee's Unleaded Blues, a juke joint on the South Side of the city. If I had a lot of work over the weekends, I'd take it with me to cafés on the North Side. I'd do my homework in a café, then do a little shopping in the thrift stores before meeting friends in one of the ethnic restaurants. Even with having to learn in a new language at a school with a reputation for tough academics, I always found time to appreciate the good things in life. That's what kept me hard at work.

At college I was exposed to real humanities classes for the first time. Fortunately, my early interest in writing proved to be an important asset. My first essay in college at the University of Chicago came back all red with an A- grade. I went to the professor and said that he should not inflate my grade because my English isn't good. I don't need pity, I'll learn. He laughed and said, "I don't care that you can't spell perfectly, can't properly use the articles and English expressions. You made a good argument, you wrote a good story, and that's what matters." I learned that I should never use "bad English" as an excuse for poor writing. Organized thoughts and well-presented arguments transcend language and subject.

In the summer after my 3rd year of college, I got to do my first real research project. By the end of it I knew I had found a home. Physics professors Sid Nagel and Tom Witten hired me to help them figure out why a drying coffee drop forms a ring around its edge. This was a perfect undergrad project, and I had a great time in the lab. I loved trying things, observing, discussing with colleagues. By the end of the summer we knew the reasons for the ring. More importantly, I had developed a taste for real scientific inquiry and the camaraderie that can develop in science labs. Even better, as I realized later, that fun summer research project about coffee drops ended up as a publication in Nature and is still more cited than any other paper to which I have contributed! Professor Nagel taught me another important lesson with that project by his insistence that the paper be well written. I learned that good communication of the results is just as important as the results themselves.

After my 3rd year at U. of C., I could have gone to grad school, but instead I chose to spend a year on a University sponsored program studying in France. The costs would be covered by my financial aid package, but the stipend was the same. At first, I cried when I realized I would have to live in Paris on a Chicago budget. What I thought was a setback turned out to be the best part of the trip. I had to move to a new city and find a way to live on a tight budget. I matured. I learned how to appreciate people who were very different than me. I lived in Paris on $30/day, and I had a blast. After the year was over, I was ready to start grad school not because that's what you do after undergrad but because I knew for sure I wanted to do scientific research.

For graduate school, I chose Princeton, where I continued to appreciate that so much important learning happens outside the classroom. Princeton was infamous for its qualifying exams known as "the generals." Princeton did not require any classes, and you gain no real credit or recognition for taking them. In order to get a Princeton Ph.D., you had to pass the generals. The generals are four days of written and oral examination on all major topics in physics – all without aids or notes of any kind. As students, we were left to our own devices to figure out how to solve problems in general relativity, condensed matter, and particle physics. Several of us grad students got together, got the books, and started studying. It was hard. It seemed kind of pointless – I wasn't going to do general relativity ever in my life. And we did it! We all passed the exam. Many years later I realized what the value of it was – we had learned how to teach ourselves, how to learn from books and from each other, how to ask the right questions and above all, we got the confidence that we were capable of figuring out pretty much any discipline. Of course, I learned a lot of physics, but the pressure and stress created a tight group out of my study partners and me. The people who were in my study group are still some of my closest friends.

The other important lesson I learned in graduate school is that choosing the right advisor is critical. When I interviewed at different schools I realized that each school has its own culture, and each lab is a unique sub-culture within the school. I realized almost immediately that certain schools and certain labs were not for me. Much like realizing that you may change majors in college, I realized that I would like to be able to choose a lab to work in after being in grad school. Your choice

41

of lab and advisor is not only a big influence on your graduate school experience, but also a big influence on your later career. Your Ph.D. mentor is like a family member; it's a relationship you have for life. Advisors are like family in other ways, too – they know you too well, you have fights, you make up. I still love my advisor, and I take every opportunity I can to go and see him. It's like visiting an old uncle.

Jobs, Family, Kids

My first jobs were at the national labs. I picked a national lab career because tenure-track scared me. I wanted to choose where to live, not just go to some place in the middle of nowhere because that's where the faculty opening was in my field that year. I wanted to have kids. Everyone said how hard it was to raise money and publish papers in the time you had to make tenure. I did not want to postpone kids until after tenure. Maybe I was not as secure as some of my friends thought. Or maybe other women are even less secure.

So, I went to a national lab, raised outside money, and published papers. Even though grad students were really hard to get at a national lab, I managed to hire and advise grad students. What I did there would probably have earned me tenure at any of the best universities. I also had two children, a boy and a girl. I took several months of maternity leave, and I wish I had taken more. Take at least four or even six months maternity leave if your finances allow. Before going on leave I was so worried that my research was going to cease to exist if I were not there. In the grand scheme of things, a few months is not that long for academic research, but your baby will never be a baby again. And, by the way, when they are less than 6 months old, they are great to travel with. The trip to Italy that I took with my five-month-old daughter was priceless and irreplaceable.

If you want to have kids, it's not which college or grad school you choose that will determine how successful you are in your career. The most important decision for your career is the person you choose to raise the kids with. Make sure that you can work well as a team with the partner you choose for life. My husband also has a successful professional career. We share responsibilities for kids, cooking, and everything; it's all teamwork. He even helped me put together this essay! Both of us have had to make sacrifices to make it work, so be sure you and your partner know how to compromise.

Be wary of the idea that two career professionals can raise children on their own. In my experience it's possible, but not desirable. Extended family and a network of friends are important in helping to raise happy, healthy kids. We've been blessed to have my parents live with us most of the year, and we are the envy of everyone we know. You will need a support network of some kind, and family is often the easiest to maintain. Because of what my parents have done, if my kids want us, my husband and I are ready. We'll move to wherever they are and help them out as much as my parents helped us.

After my younger kid was over one year old, I decided to take another challenge. I started a company based on the research my group at the national lab had done. That's where I am now, and I couldn't be happier. I have had to learn accounting and government contracting and pass defense department audits. I know more than I ever thought I would about water markets. I manage almost 20 people and many millions of dollars. These days I don't go to lunches and dinners with science geeks but instead with business guys. I've had to learn how to network. I work a lot but love the freedom. I most enjoy working with people who are really excited about making something work for real and not just for a paper.

Some Thoughts

In this narrative about my life, you'll notice certain recurring themes: don't just focus on your nerd skills, don't just focus on academics, take opportunities to travel, don't forget that you have to enjoy your life to know what it's worth. Being good at math and science is great, but you'll need other skills, like writing, to get ahead. Use your education not just to acquire facts, but to learn how to learn and how to think critically. Take time to travel, study abroad, and work in a different place. Don't just go to a math camp, but go hang out with people who are very different from you. You will grow as a person. Don't do it because it's another box you want to check on your resume. Do it because you will be a happier and better person. It's worth it; trust me. When you're struggling to finish grad school it helps to remember why you went to grad school in the first place.

For women, above all, you need to have confidence in yourself. I am a woman who left home as a teenager to live halfway around the world while war broke out near my home. I got a Ph.D. in physics in a demanding top-tier program, made my own research group in a national lab, and started my own company. I still struggle daily with the idea that I need approval and consensus to do anything. My male colleagues, even ones who I firmly believe are less capable, don't seem to suffer from these same doubts. Insecurity is a terrible thing. Mistakes that I made in life were almost all driven by some sort of insecurity. When you doubt yourself, you've got to take a deep breath and take the challenge. You can do it. And don't immediately blame yourself for something that went wrong. We ladies tend to do that more than the guys do.

As I am further along now, I try to give back some of the help I've received. I've so far had a pretty successful career and life, but it wasn't all my own doing. So many people have helped me out along the way. My parents sacrificed to give me better opportunities. RSI/CEE folks helped me get into colleges. John and Agnes Finn took me in to live with them and their family for the summer before college, so I could get out of Yugoslavia where the civil war was raging. My friends in Chicago helped me get by in the new and very unfamiliar place when I came all by myself. University of Chicago paid for my education and for me to enjoy Paris. My husband, well, he's still helping with everything.

DANIELA BORTOLETTO

Bona Fides

Dr. Daniela Bortoletto graduated summa cum laude from L'Università degli Studi di Pavia with a Bachelor's degree in Physics. She received her M.S. and Ph.D. from Syracuse University. She completed her postdoctoral fellowship at Purdue University where she is currently the E. M. Purcell Distinguished Professor of Physics. She is a Fellow of the American Physical Society, and she has received several national awards including the A. P. Sloan Fellowship and the National Science Foundation Career Award.

Into the Woods

I am surprised that I became a physics major, traveled to the U.S. to receive a Ph.D. in physics, and then became a physics professor in a major U.S. university. This was a very unlikely trajectory, since I come from a beautiful small town in the Italian Alps called Domodossola, and I am the first member of my family to even get an undergraduate education. I was always an avid reader, and when I was about twelve I became interested in physics after reading a book by George Gamow. He was a great storyteller who was able to convey great excitement about past revolutions in physics like quantum mechanics and relativity.

My father died when I was only 9 years old, leaving my mother and I in a rather shaky economic situation. When I finished middle school my mother encouraged me to enter a training school in accounting, which would give me a diploma without attending university, and very likely lead to a well-paying job. To attend university at that time one had to complete a five-year-long preparatory program called "liceo." In my hometown the only "liceo" was a very expensive private school that was clearly out reach for my family. My first lucky break came in 1972, just when I was to enroll in the first year of high school, when a public "liceo" opened in a neighboring town; the school moved to Domodossola the following year. After a struggle, my mother agreed that I could attend the public "liceo," even if in the end I would not receive a diploma and would have to attend university.

As a high school student I excelled in every subject, especially physics and mathematics. In the final exams of the last year I was one of two students in my hometown to get the top mark of 60/60. I was quite uncertain about which degree I would seek after finishing high school. I originally planned to enroll in "chemistry & pharmaceutical technology," a degree that at the time was believed to lead to a good job. It was my physics professor, Mauro Magri, who first told me that my mathematical skills were very high and that I should enroll in mathematics or physics. I chose physics at the University of Pavia, and I have never regretted it.

The University of Pavia is one of the oldest universities in Europe. Teaching in Pavia dates back to around AD 825, originally focusing on ecclesiastical and civil law and divinity. The university was officially established in 1361 by Holy Roman emperor Charles IV. Pavia is a charming university city, and it hosts three famous "colleges," cultural institutes which provide support to high-achieving students.

My second lucky break came when I was admitted to Collegio Ghislieri. The "collegio" was founded in 1567 by Pope Pius V to accommodate promising students experiencing economic hardship. Promising women were admitted starting in 1965. The "collegio" selects students of the University of Pavia through a national competition based on a rigorous written and oral examination. The students who are admitted every year are required to pay a contribution that is proportional to the family income, which was a great relief for my family. The services offered by the college are not limited to food and housing but focus also on providing further training. It was a great growing experience for me. Only five physics students were admitted in my year, but I was no longer the smartest person in my class.

The CERN Summer Student Program

The environment at the University of Pavia and the competitive atmosphere of the "collegio" were extremely important to me. However, being admitted to the CERN summer student program really changed my life. Being at CERN was a mind-blowing experience. I had fabulous teachers. The research atmosphere was incredibly intense, and I decided there that I should become a particle physicist. I also had a lot of fun and met many young people from all over the world who shared my passion for physics.

I went to CERN in 1980, and some of my summer student fellows were working in UA1 and UA2 experiments. In 1979, CERN converted the Super Proton Synchrotron (SPS) into a proton–antiproton collider using stochastic cooling to accumulate sufficient numbers of antiprotons to make a beam. This technological advance, invented by Simon van der Meer, led to the discovery of the W and Z particles in 1983 by UA1 and UA2. This was a huge discovery, and Carlo Rubbia and Simon van der Meer received the Nobel Prize in physics only a year later.

The Tale of Two Worlds

After completing my undergraduate degree at the University of

Pavia I decided that I should go to the USA for my Ph.D. The same choice was made by a few of my fellow summer students. My advisor Sergio Ratti also encouraged me to make this choice since the Italian physics Ph.D. program was just starting and was not yet well established. An Italian professor, Giancarlo Moneti, working at Syracuse University, came to Pavia and gave an interesting talk about an experiment in the USA called CLEO. I was able to meet Marv Goldberg, another professor from the same university, at CERN. They both encouraged me to apply to the Syracuse University graduate program, which I did, and I was accepted as a research assistant in particle physics.

Though I was excited to explore a new country, I was very sad to leave Europe since just before leaving for the U.S. I had met a handsome British Ph.D. student named Ian Shipsey. It all worked out in the end however, as he is now my husband.

I was immediately impressed with the U.S. research system, which seemed more dynamic, democratic, and less formal than the one in Europe. For example, I was the student representative during a search for a new faculty member, and I was invited to meet all faculty candidates. Nothing like that would ever happen in Italy.

I was also surprised by the fact that there were practically no female students in the Ph.D. program. My Italian experience was very different. For example, in Collegio Ghislieri, out of the 5 students admitted in physics 3 were women and 2 were men. Furthermore, every time I met somebody and they learned that I was studying physics, they seemed surprised that a woman would choose this subject. My interest in trying to increase the representation of women in physics started at that point. It appeared that in the U.S. many people believed that women were not capable enough to be physicists. I knew that they were wrong since there were many women physicists in Italy.

Discoveries in Particle Physics

I joined the CLEO experiment for my Ph.D. studies. The collaboration contained about 100 physicists and was dedicated to the study of the b-quark. This quark, which was predicted to exist in 1973 by Kobayashi and Maskawa to explain CP violation, had been discovered in 1977 at Fermilab. The detailed study of the properties of the b-quark was carried out mainly at two experiments, CLEO and ARGUS, an experiment at the DESY laboratory in Germany. My thesis provided a measurement of the parameters of the Cabibbo, Kobayashi and Maskawa matrix element called V_{cb}, which measured how often a b-quark changes its flavor and becomes a c-quark.

It was very interesting to work in CLEO since a student could play a large role both in the construction of the apparatus and also in data analysis. For example, I participated in the construction of the CLEO muon chambers, and my analyses led to 5 physics papers. It was just fantastic.

My husband-to-be, Ian Shipsey, also joined CLEO as a postdoc, and we got married in 1988. Unfortunately, just 6 months after our marriage and before I graduated, Ian got very seriously ill with leukemia. It was quite a challenging time, but with the help of my advisors Giancarlo Moneti and Sheldon Stone, one of the world experts in b-quark physics, I was able to focus and obtain my Ph.D. in 1989.

Ian recovered almost miraculously, and we started to coordinate our careers in physics very carefully. Nearly 45% of married women physicists have physicist spouses and therefore face "dual-career" issues, since finding two positions at the same university is often difficult. Ian accepted a position as an assistant professor in particle physics at Purdue University, and I joined the same university as a postdoc. I remained in CLEO for three years as a postdoc, and then I started to apply for faculty positions. Since I received offers from other strong institutions, Purdue came through with an assistant professor position. At that point I decided to move my research to the CDF experiment at the Tevatron proton-antiproton collider at Fermilab in Chicago and to start exploring the energy frontier.

CDF was a great experiment, and in 1995 I participated in the discovery of the last quark predicted by the standard model, the top quark, a 172 GeV elementary point-like particle almost as massive as a gold atom. I also started to work on silicon detectors for precision measurements of the positions and trajectories of charged particles. The high granularity provided by silicon detectors has allowed precision measurements at hadron colliders. My experience in CDF and the Tevatron was perfect training for my current experiment, CMS at the Large Hadron Collider (LHC).

I started to work on CMS in 1996, and I participated in the construction of the Forward Pixel detector. Freeman Dyson said: "New directions in science are launched by new tools much more often than by new concepts." I believe that the pixel detectors now installed in ATLAS and CMS have had a major impact on the performance of the LHC detectors and their success. This success culminated this summer with the discovery of a fundamental scalar with properties consistent with those of the long-sought Higgs boson.

As it is often said "Extraordinary discoveries need extraordinary scrutiny." To avoid any possible bias while analyzing the data to search for the Higgs, every CMS analysis group draw "blinds" over the region where an excess of Higgs events is expected. Part of my group works on the Higgs to ZZ to 4l channel (4l means two pairs of electrons, or two pairs of muons, or one pair of electrons and one pair of muons). It is one of the channels most sensitive to the Higgs. The 4l group was asked to unblind the data on June 14, 2012 so that the analysis could be discussed internally on June 15 by the Higgs group and give enough time for the entire collaboration to review it and present it at a public seminar at CERN scheduled for July 4, 2012. The "unblinding" of the data was one of the most magnificent moments of my life. It was clear that we had finally made a major discovery, and those years of hard work were finally paying off. Many women shared this exhilarating moment. Some were in the room, and some were participating in the meeting through a video connection. Some, like my colleague Chiara Mariotti, had a major impact not only in this discovery but also in educating the next generation of women participating in this work. Only further studies will tell us if this particle with mass of 125 GeV is in fact the Higgs; nonetheless we know that this discovery is likely to guide us in the path to find a more fundamental theory of particles and their interactions. I am sure that the excitement brought by this discovery will bring a new generation, and hopefully many women, to the study of physics.

Reflections and Recommendations

Because I started my training in Italy, where there are many women in physics, I never had any doubt that women are as capable as men in this field. My advice and recommendation is to follow your passion, believe in your capabilities, but also to build a support structure around you.

My family is very important to me. I visit my mother as much as I can; she still lives in Italy. My husband is a great companion and a fellow physics explorer. Finally, I admire and love my daughter Francesca, who has a great passion for music and wants to become an opera singer.

A career in physics can be challenging for both men and women. Every particle physicist has to be able to work in large collaborations, at experiments often located great distances from home. This can be very stressful. What has helped me the most in my career is to look back at what motivated me in the first place to enter this field. At the core of my passion is the possibility to be among the first to unveil what is produced at higher and higher energies as we continue to probe the constituents of matter and explain the nature of our universe.

PATRICIA R. BURCHAT

Bona Fides

Dr. Patricia R. Burchat received a Bachelor of Applied Science and Engineering in Engineering Science from the University of Toronto and a Ph.D. in Physics from Stanford University. She was a postdoctoral fellow with the Santa Cruz Institute for Particle Physics and then a professor at the University of California at Santa Cruz, before returning to Stanford University where she is the Gabilan Professor in Physics and the Sapp Family University Fellow in Undergraduate Education. She has served as Chair of the Physics Department and has held leadership and management positions in the BABAR particle physics collaboration, the Large Synoptic Survey Telescope collaboration and the Kavli Institute for Particle Astrophysics and Cosmology. As an Assistant Professor, she was awarded a National Science Foundation Presidential Young Investigator Award. She is a Fellow of the American Physical Society and is a Guggenheim Fellow. At Stanford, she has been awarded the Dean's Award for Distinguished Teaching and the Walter J. Gores Award for Excellence in Teaching. She has served on the Committee on the Status of Women in Physics in the American Physical Society and has been a leader in the organization of the Conferences for Undergraduate Women in Physics. Pat was elected to the American Association for the Advancement of Science, and she is the recipient of the 2010 Judith Pool Award for mentoring young women in science.

PATRICIA R. BURCHAT

From the North Woods of Canada...

I grew up in Barry's Bay, Ontario – a town of 1,200 with an economy based on logging and farming.

The first regional high school in Barry's Bay was built in 1967, Canada's Centennial year, when I was nine years old. The population of the high school peaked while I was there at around 800 students, many of whom rode to school on buses from the surrounding wooded and rural area. The construction of the high school granted my eight siblings and me the opportunity to become first-generation high school graduates and pursue higher education.

My homeroom in high school happened to be the drafting classroom. I was fascinated with the technical drawings on the walls and soon signed up for a drafting elective. When a small land-surveying office opened in town, the owner asked the drafting teacher to recommend a student to run calculations and draft survey plans on weekends and during the summer. I don't think the owner was expecting the skinny girl who showed up for the job. This work was a huge step up from my job at the local dairy/coffee shop/bus depot. I really enjoyed the summers I spent drafting plans with ink on waxed linen (no mistakes allowed!) and programming some of the first programmable calculators.

I had a number of very good math and science teachers in high school. In Grade 13 – yes, we had 13 grades in Ontario at that time – I began to pour over the university course bulletins in the guidance office. Although I studied one year of physical sciences and two full years of physics in high school, it did not occur to me to pursue a degree in physics. What would I do with a physics degree? I concentrated on the engineering sections of the university handbooks and came across a sentence in the description of the engineering science major at the University of Toronto: "Only those with above average abilities in math and physics should apply." I accepted the challenge and applied to this elite limited-enrollment program. Back then in Ontario, applying to university meant prioritizing the list of universities and majors on a form and mailing it in. The high school sent the transcript and that was it.

... To Engineering Science

The engineering science major turned out to be a perfect gateway to the world of higher education. It placed a strong emphasis on understanding concepts in math, science and engineering from first principles, and the application of these concepts to identifying and addressing engineering challenges and outstanding questions in science. Last year, I worked with the School of Engineering at Stanford to introduce a similar major: engineering physics. One of my stated motivations for proposing this new major reflects my own experience:

> "To many students and parents the 'security' of a degree in engineering is very important. For some students in the School of Engineering, a physics major would be a very natural fit to their talents and interests. However, some of these students may be reluctant to pursue a degree in physics because the career options appear to be less well-defined. This could be particularly true for first-generation college students. For these students, an engineering physics degree can provide a path to higher education in engineering or physics."

My engineering science class at the University of Toronto contained many first-generation immigrants and first-in-family college attendees – a large fraction of whom went on for higher degrees. I recently returned to the University of Toronto – 30 years after graduating – to review the engineering science program. I was delighted to see that the fraction of women in the program had increased by a significant factor since the time I was there.

Using Your Summers Wisely

One of the pieces of advice that I give to my undergraduate advisees at Stanford is to use their summers wisely. As an undergraduate, you have three valuable summer opportunities to learn what you might enjoy doing and – as importantly – what you would not enjoy doing beyond a summer experience. One summer I worked for Shell Oil where I had some exposure to chemical engineering; after this experience, I decided that the chemical engineering option of engineering science was not for me. I decided to pursue the nuclear and thermal power option (an option that no longer exists). My main motivation was to understand whether the heavy water "CANDU" reactors used in Canada were safer than the light water reactors used in most of the rest of the world. Since then, I have continued to have an interest in energy issues.

In the summer after my junior year, I worked as a research assistant at the Chalk River Nuclear Laboratories. The lab was built during World War II in a beautiful but isolated location on the Ottawa River, only about 50 miles "as the crow flies" from Barry's Bay. This summer experience opened the world of research and scientific discovery to me. I had fantastic mentors that I aspire to model in my own research group today. The following summer, I worked at the Fermi National Accelerator Laboratory near Chicago, before coming to Stanford University for graduate studies in experimental particle physics. By this time, I had realized that I was a "reductionist" at heart. I am fascinated by fields of study in which one tries to understand nature at its most fundamental level, whether that be at the most microscopic level, through the investigation of the elementary particles and the fundamental interactions, or at the most macroscopic level, through studies of the contents, evolution, and ultimate fate of the universe.

The Academic Career

After graduating with a Ph.D. in experimental particle physics from Stanford, I worked as a postdoctoral research associate and then an assistant and associate professor at the University of California at Santa Cruz. By this time, I was married and had two children. We lived in Palo Alto. My husband commuted an hour to San Francisco, and I commuted an hour in the opposite direction, twice a week, to Santa Cruz – not an optimal situation. At first I was reluctant to even consider leaving UC Santa Cruz for Stanford; the role of the public universities was and is very dear to my heart (all universities in Canada are public). However, around that time the UC Board of Regents was considering barring the use of affirmative action in university admissions, while Stanford was affirming its dedication to educating a body of students that reflects the population it will lead.

I came back to Stanford in 1995 and have enjoyed my time here immensely. I receive great satisfaction from teaching, mentoring students, and being an active member of the broader university community. I enjoy my interactions in research with colleagues in particle physics, astrophysics and cosmology from all over the world. I also enjoy leadership and management roles that allow me to help bring about positive change. I have found that most people respond well to an energetic leader with a positive attitude, who is prepared to listen – but does not tolerate petty politics.

One of my passions is to help college students thrive in technical fields, even if their high school did not offer good learning opportunities in math and physics. I thoroughly enjoy teaching large introductory physics courses, engaging and training graduate teaching assistants to help every student succeed, and instilling a life-long, positive view of physics in all my students.

One Step at a Time

Throughout my path from a small town in Canada to chair of

physics at Stanford, I never really thought beyond the next step. I likewise advise my students to take things one step at a time – what would you most like to do next? What feels right? For one of my advisees who majored in physics and minored in materials science, it was taking a "gap year" to help restore a national park in Patagonia; she entered a Ph.D. program in applied physics afterward. For another physics major who minored in creative writing and poetry, it was teaching English in Japan for a couple of years; he then entered a Ph.D. program in physics. I encourage you to follow suit and do what feels right at the moment. Your path will find you.

CATHRYN CARSON

Bona Fides

Dr. Cathryn Carson studied physics, mathematics, and the history and philosophy of science at the University of Chicago. Her graduate work took place at Harvard, where she received a master's degree in Physics and a doctorate in the History of Science. After a postdoctoral year at Stanford, she joined the history department at the University of California, Berkeley. Her research focuses on the history of physics and philosophy in twentieth-century Germany and the United States. Most recently, she published <u>Heisenberg in the Atomic Age: Science and the Public Sphere</u>. She chairs the editorial board of the journal <u>Historical Studies in the Natural Sciences</u> and serves as Associate Dean of Social Sciences, the largest division at Berkeley.

Trajectories

My story is different because I am not a physicist. Rather, I passed through physics on my way to becoming a historian. As a historian, narrative is one of my tools, and so when I tell stories, I can't help but be conscious of how they are constructed and told. My story as a "woman in physics" is the narrative as I make sense of it now, meaning as I pick out an origin and a trajectory that leads to the present.

I am part of the generation that went through school and

college in the 1970s and 1980s, then graduate school in the early 1990s. Throughout this time I was hyper-aware of being a woman in science. Even though the formal barriers to women in physics were gone, the informal social relations around them were still in flux. In its own backhanded way, that was productive for my intellectual development, as I am someone who studies the social relations of science. I feel fortunate to have made the passage through physics, and at the same time I'm glad to have found something on the other side.

Steering into Math and Science

I don't know how early a child can become aware of gender roles. I have distinct recollections from nursery school of feeling that girls got the short end of the stick. Girls were supposed to be well-behaved and polite and play with dolls while boys got to run around. I have no idea where I got the sense that some people felt girls were naturally suited to doing only certain kinds of things. It was not from my family. My parents were completely sympathetic to a tomboy daughter; my father coached my soccer team back when a girls' league was still something new. My mother has a Ph.D. in economics, and my sister has a J.D. We are all, however, well-behaved and polite.

I first became existentially aware that gender intersected with science when I was part of a large research project called the Study of Mathematically Precocious Youth (SMPY). The core of the study back then was to have seventh graders sign up for a college entrance exam, the Scholastic Aptitude Test. I didn't realize there would be an awards ceremony after the results came out. I was completely surprised to be invited up on stage, together with another girl, to recognize our twelve-year-old SAT-verbal scores. Then I sat in my seat, my surprise turning to I didn't know what, as the researchers called up the other high scorers. Every seventh-grader that year who scored above 700 on the math section was a boy, and the researchers seemed excited, in fact determined, to drill this finding into the minds of the assembled students and parents.

A twelve-year-old's fury can take you a long way. I was close enough to the math cutoff that I left the auditorium feeling I had failed. I don't know that I told anyone, but I was not going to let that happen again. It was a twelve-year-old's black-and-white view of the world, but it took hold at a moment when long-term trajectories were being set. SMPY offered accelerated summer math classes, and I took them. I learned to code in BASIC and Pascal and willed myself to play ultimate frisbee and juggle and do other things that mathematically gifted kids (boys) did. In the confusion of wanting to know my life's destiny, I decided that destiny was science. And not just science, but physics. Theoretical physics was the hardest thing you could do. Coupling fury with stubbornness and competitive ambition made for a considerable force – even when internalized social norms made it impossible to own those traits and express them out loud.

I was incredibly lucky to have great teachers in high school, a private girls' school. It felt supplementary at the time, but I got real preparation in history, literature, and languages. Before graduating I was able to take single and multivariable calculus. I took AP Chemistry and AP Physics in the company of some equally ambitious young women – and we took these classes at the boys' school down the street. Gender was always part of the picture. The first U.S. team for the International Physics Olympiad was chosen the summer I graduated from high school. I still have a commemorative photo from the training camp: the participants posing on the massive statue of Einstein outside the National Academy of Sciences, eighteen boys in suit jackets or shirtsleeves, two girls in a dress and a skirt with summer-white shoes.

Living in Physics

What I loved about physics, to start, was quantum mechanics, at least as far as it filtered through to an actual student of physics. It was easy to be pulled in by the abstract clarity of classical mechanics, and when I first encountered the connection between Poisson brackets and canonical commutation relations, it was like some inner structure of the world was revealed. I was fascinated by turbulence and scaling laws, and I found a hard-to-describe pleasure in thinking with the tools of condensed matter theory. The frustration that had set in in high school, and only deepened as I settled into serious study, was just that a lot of the day-to-day work of physics was not really thrilling. Practically, physics came down to endless iterations of boundary value problems in electricity and magnetism, or painfully long solid-state problem sets worked out on rolls of dorm paper towels. In E&M lab sections I would propagate all the measurement errors while my partner made the equipment work, and I realized I was much more interested in rewriting the lab manuals than in learning the handiwork they were supposed to be teaching.

Being a woman in physics didn't make this harder; it probably made me stick with it longer than I would have otherwise. When I declared my major as the history and philosophy of science (to my great good fortune, Chicago had this option, and I stumbled on it), it felt like a declaration of independence. The fact that I avoided majoring in physics probably made sense to my undergraduate friends and the department staff who heard me muttering about how ambivalent I was. But I kept taking the classes to satisfy all the requirements, and as long as you do the work for a degree in physics (Chicago didn't allow double or triple majors back then), you might as well do the work for a degree in math, too. I had some degree of awareness about the bind I was putting myself in. It was just easier to be reflective in sociological terms than personal ones: these are the dilemmas marginal groups face, convoluted with this is how upper-middle-class white women behave. If I stopped taking physics, I would have had to really figure out what I wanted. I would also have felt like I was letting down the cause.

As an undergraduate I had only occasional experiences of feeling out of place as a woman. There was the lab tech at a university in the Southeast who told me, smilingly, that he didn't think women belonged in physics, which didn't trouble me, because I was visiting to do research (I eventually published two papers in computational condensed matter) while he was building bookcases with his degree. In graduate school it was harder to avoid. That was not because of any hostility I picked up on, but because of the starker difference of perspectives, and because of the numbers narrowing down. There were male students, friends, even, who didn't get that it might feel odd to be the only woman in a condensed matter graduate seminar. (At that point I was already a refugee from another department; I had decided to go for my Ph.D. in the history of science.) I still have the old flyer – I was thrilled when it went up, and I quietly took down a copy and saved it – announcing the party to celebrate Melissa Franklin's tenure, the first woman in the history of the Harvard department.

I was glad for the chance to watch that history happen, and to have a sideways foothold in the department as it did. And still it was gendered. A wonderful mentor pointed me to Evelyn Fox Keller's reflections on her much earlier experience at Harvard. A lot of things about social relations had changed, and then others seemed just the same. To someone with a background in feminist theory, the department's "family" meetings seemed like sites of gender-reversed parthenogenetic self-reproduction in the ironic absence of women. A friendly person on the third floor of Jefferson Lab pointed out it was easy to tell when I was coming around the corner because other grad students wore sneakers and I wore clickety flats.

Passing Through

In graduate school the understanding slowly came to me that my mind worked differently from those of the grad students who had offices in Jefferson Lab. They didn't remember formulas as strings of syllables; they didn't live in language the way I did. Running up against my limits – looking for answers to problem sets in the old volumes of journals in the storage room – let me admit that this didn't have to be my calling. Fortunately, I had already found my way to something I cared about, thinking about human affairs unfolding in time.

I was trained as a historian, specifically of physics, understood as a human affair unfolding in time. That is now what I teach, and part of what I do. As I have spent more time as a historian, I have tried to line up my "internal" interests in how the science developed with questions of understanding human choices and institutional structures. I get mildly frustrated when I bump up against some scientists' assumption that being a historian (or a humanist or social scientist in general) is somehow soft (and therefore coded female). There is a critical rigor that animates some work in the "soft" fields, certainly my own, exactly in how it deals overtly with the limitations of knowledge, perspective, contingency, and method that the natural sciences don't have to address.

My husband, who was my E&M lab partner, says it makes sense that the possibility of recognizing structure in turbulence would appeal to someone who was cut out to study the complexity of human behavior. Condensed matter metaphors underwrite how I think about history, and the ways of parsing reality that physics teaches have become a kind of conceptual shorthand. I don't use my physics training anymore, except occasionally to put people on notice. When I work with quantitative social scientists (or scientists and engineers) it helps diffuse the unspoken skepticism that a woman and a historian sometimes encounters. I once put it this way to a colleague in engineering: having trained in theoretical physics doesn't mean I know everything, but it means I'm confident there's nothing I can't learn. That's not true for the reasons physicists may think – because physics can handle everything, or because physics is the hardest thing to learn – but it's become true for me.

Reflections

I now talk about my encounter with physics with a historian's distance. I tell it as a larger story about social and cultural transformations in the United States after World War II. As the nation geared up for an era of massive international competition, it invested all-out in scientific and technical training. Students with those aptitudes were seized upon as gifted and steered into professions that gained a cultural premium they had never had before. Women and other marginal groups were only slowly added to the human resource pool; that happened fairly late in the game. But we are still dealing with the 1950s tinges to our notion of a scientific career. These don't just impinge on the way we think about individual trajectories. They're also built into pipeline models that integrate over individual choices to give some idealized lossless flow.

This system was incredibly effective in recruiting physicists. It did not exactly adjust, though, to the complexities of the people who were fit into the mold. And it was not respectful or attentive to other things an individual might find fascinating. The smart, ambitious women from my AP Physics class are now working on Wall Street, practicing law, editing, and teaching and writing about comparative literature and history. I can appreciate the frustration other women have felt trying to stay in physics, but I can't think of passing through it and leaving as solely a bad thing.

SHIRLEY CHIANG

Bona Fides

Dr. Shirley Chiang received her A.B. degree summa cum laude in Physics from Harvard University and her M.S. and Ph.D. degrees in Physics from the University of California, Berkeley. She was a Research Staff Member at the IBM Almaden Research Center for eleven years before joining the faculty at the University of California Davis as a Professor. She is an experimental condensed matter physicist who specializes in using high resolution microscopy to image solid surfaces with atomic and molecular adsorbates that are involved in chemical reactions, nucleation and growth phenomena in metal-on-metal and metal-on-semiconductor epitaxy, and surface structural phase transitions. Shirley won the Outstanding Mentor Award from the Consortium for Women and Research at UC Davis and the UC Davis Distinguished Teaching Award. She served as Chair of the Department of Physics at UC Davis for five years and is a Fellow of the American Physical Society (APS), the American Vacuum Society (AVS), and the American Association for the Advancement of Science (AAAS).

Early Years

I was born in the United States as the elder daughter of Chinese immigrant parents who had both come to this country with student visas to study economics. My father had been working for the Central Bank of China and got the opportunity

66

to obtain a graduate degree at the University of Chicago. After the establishment of the People's Republic of China in 1949, he finished his Ph.D. degree and stayed in the U.S., meeting my mother in Chicago several years later. My parents always emphasized the importance of education, as their passing exams and attaining good grades had enabled them to come to this country, which they valued as the land of opportunity. My father had a long career as a university professor. My mother, however, always regretted that she did not get also get a Ph.D. degree because she left graduate school after I was born. Later, she became an early computer programmer and eventually got an M.S. degree in applied mathematics in night school, which took many years. Thus, I later became determined to finish my Ph.D. degree before getting married.

I was always the top student in my class throughout elementary and secondary school. Although I always got excellent grades in mathematics and science, I remember my father being particularly proud when I got the highest grades in English, as his own English writing always seemed less than perfect. We were the only Chinese family in the northeastern Pennsylvania city where I spent most of my first ten years and one of a handful when we moved to a suburb of Rochester, New York, where I finished high school. My parents had the idea that "smart" people were scientists, and thus they bought me all sorts of science kits when I was a child. In grade school, I remember building a small telegraph assembly and an electric motor. I became a voracious reader and was particularly impressed by biographies of Marie Curie, the Polish-French scientist who won Nobel Prizes in both physics and chemistry, and Albert Einstein. Nevertheless, I still had no real idea for many years about what physicists study or do.

When I studied chemistry in high school, I first thought that it was the most interesting subject I had seen. Thus, the next year, I studied Advanced Placement (AP) chemistry concurrently with physics. Although my score on our first physics exam was terrible, it was still the highest one in the class. Fortunately, our high school physics teacher had a system of retests, and I attained the highest possible score on the first retest and never needed to take another one in that class. I then became much more interested in physics than chemistry, since we could use equations to explain and predict physical phenomena.

The following summer, I was fortunate to be chosen for an NSF physics program at Cornell University. There we studied some special relativity and quantum mechanics, while dabbling with electronics in the laboratory and going to special lectures about current physics research. Particle physics seemed particularly exotic and interesting. I was especially impressed by a description of the plot of mass versus the third component of isotopic spin for a set of particles, which led to the prediction of the existence of the Ω^- particle and its subsequent discovery.

I went back to high school for my senior year and promptly filled out all of my college applications saying that I wanted to be a physicist. I recall having an interview at Radcliffe College (admissions were separate from Harvard then) during which the woman interviewer, who clearly knew nothing about physics or how excellent the Harvard department is, told me that if I was so interested in physics, I should consider going to MIT.

University Years as a Student

One reason I decided to go to Harvard instead of MIT was that Harvard gave me sophomore standing upon entrance because I had many AP credits. That meant, however, that I needed to figure out how to take all of the required physics courses in three years. I talked my way into Physics 55 (the one-year course on electricity and magnetism and electromagnetic waves for advanced freshmen) and rapidly found out that I was not nearly as well prepared for this course as most of my classmates, who were all men except for one other woman. By working very hard, I managed to stumble through the culture shock of freshman year, finally feeling like I had gotten caught up by the end of the year.

After my first year at college, I got a summer job at the Kodak Industrial Laboratory, where they did chemical analysis of materials from the film manufacturing areas, such as silver in film processing chemicals and dirt collected on paper filters from air samples. While rolling up paper filters for examination in a spark spectrograph was tedious work, writing a computer program in FORTRAN to analyze the data was more interesting.

Back at Harvard, I took the junior year physics and math classes, including electronics and introductory solid state physics. I found this last class particularly boring because the famous professor did not put the many topics into context and regularly put me to sleep in early morning lectures by talking in a monotone. Nevertheless, taking this course got me to the attention of Gunther Wertheim, a scientist at Bell Laboratories who had gotten his Ph.D. degree at Harvard; he became my next summer research supervisor. This was a real turning point in my life, as through that job, I became interested in studying solid surfaces, the field in which I still do research today. In addition, I found out that I like experimental physics and being able to control my own experiments in my laboratory. When Bell Laboratories subsequently also gave me a yearly grant for graduate school, I was fortunate to be able to work with several more prominent condensed matter experimentalists there, particularly Neville Smith, who served as a mentor for many years. Neville took me to the University of Wisconsin at Madison to work for three weeks on photoemission experiments at the small dedicated Tantalus synchrotron radiation ring. I recall his being very embarrassed when explaining to me that I was the first woman to work there and that they had only one restroom; they put a reversible sign on the door so that I could use it also.

After starting graduate school at UC Berkeley, I was particularly interested in looking for a physics research project on surfaces. I found one in Paul Richards' group, where I was given the opportunity to build a low temperature infrared spectrometer to do the ambitious experiment of measuring the infrared emission signal from adsorbed molecules on a metal surface at room temperature. Paul's group was unusual in that it was tied together by the common use of infrared spectroscopy, while the students worked on diverse scientific problems ranging from surfaces and infrared properties of materials, to superconducting and semiconducting infrared detectors, to infrared sky surveys and measurements of the cosmic background radiation. Paul not only taught me how to build equipment but also how to think about the limits of signal to noise in measurements.

Industrial Research

After finishing graduate school, I was interested in an industrial research job for numerous reasons. My experience at Bell Laboratories convinced me that industrial research could be both very interesting and rewarding. Also, I enjoyed and wanted to continue doing the experiments myself and not just to supervise other people doing research. I was fortunate to obtain a permanent staff position at the IBM San Jose Research Laboratory (later renamed Almaden Research Center) without postdoctoral experience. Not only did this job pay better than an academic postdoctoral position, but it was also relatively easy to get research funding inside the company. In addition, since I was by then planning to get married, it was attractive that we would only need to solve the "two-body job problem" once. At that time, people tended to go to IBM for a whole career; plus Silicon Valley had many companies, and therefore, multiple job opportunities.

I first heard about the scanning tunneling microscope (STM), which could resolve individual atoms on a solid surface, from an invited talk given by Heini Rohrer from the IBM Zurich

Research Laboratory at the APS March Meeting in Los Angeles in 1983. I was therefore thrilled to get the job to build an ultrahigh vacuum (UHV) STM at IBM Research in San Jose, in collaboration with Robert Wilson, who had been hired that same year. I got married and visited the IBM Zurich Laboratory briefly while we were on our honeymoon in Europe.

The first STM conference was held in Oberlech, Austria during the summer of 1985 and was attended by about 35 early adopters of the instrument. I still remember that Heini Rohrer opened the conference by saying, "Gentlemen and Shirley..." Once we finally got our instrument working, it was an exciting time because almost every material we examined with it gave us new and unexpected results. We made the first images of close-packed Au atoms on a surface, high resolution images of metal-on-semiconductor ordered structures, and images of individual metal atoms in ordered alloy layers. Our measurement of individual benzene molecules on a rhodium surface was even published in the New York Times. I also did experimental and theoretical images of a series related molecules on a platinum surface with a woman chemist, Vickie Hallmark, who became a very close friend. Together with another colleague, Gary McClelland, and two talented postdoctoral fellows, Ragnar Erlandsson and Mathew Mate, I built an early atomic force microscope, which we used to make the first measurements of atomic scale frictional forces on graphite.

By the early 1990's, however, corporate research at IBM was changing, as management wanted us to make our research more relevant to issues in the manufacturing plant very quickly, without giving us additional money for new laboratory equipment. For a period, lunchtime conversation was dominated by the discussions of who was planning to take the next early retirement offer. I did have the opportunity to work for a year on the development of a very different type of surface microscope, a spin-polarized low energy electron

71

microscope (LEEM), which allows the imaging in movies of surface magnetic properties with lateral resolution of 10nm.

So I began to consider going back to academia. Because I had an excellent publication record at IBM, I was able to obtain a full professorship at UC Davis, a place which was particularly attractive because its reputation was definitely going up, in contrast to IBM.

Back to the University as a Faculty Member

I found that I love working at the university, including all aspects of the multifaceted work of a faculty member: research, teaching, and service. I particularly enjoy teaching classes and working with students in the laboratory. I now enjoy teaching other people how to do things in the lab and no longer want to do everything myself. Working with talented students is very rewarding, and it is still fun to obtain great data and think hard about the interpretations. It is also wonderful to see them grow in their abilities and understanding and then to establish their own careers.

Since I definitely had more than my share of women graduate students during my first few years at the university, I assume that they liked having a woman role model, something which I had never had. Although I had been fortunate to have several mentors through graduate school, they were all men. In addition, about half of the forty undergraduate researchers who have worked in my laboratory have been women. I subsequently received a mentoring award from the UC Davis Consortium for Women and Research.

I had told the vice chair in charge of teaching assignments that I might be interested in teaching graduate electricity and magnetism "someday;" that day arrived sooner than I expected, and I found myself relearning applied mathematics in order to teach it. The first year I taught those courses was the year when the third edition of Jackson's textbook was published, and I had to rewrite all of my problem solutions in SI units! The students and I worked through many difficult problems together in long impromptu office hours. Their support and letters led in large part to my receiving a university teaching award.

I even enjoy committee service, both at the university and professional levels. I probably do too much of it, however, as women are often in demand on committees, and I have a hard time saying no to requests for help. As department chair for five years, I was a good administrator of a large department and was able to get consensus among department members, even on controversial issues.

One of a Few Women

Over the years, I have become inured to being one of a few women in both physics and academics. Most of my colleagues and friends are men. When I have actually had the opportunity to work with a woman colleague, we have become close friends.

At Harvard, I remember attending one meeting of undergraduate women physics students. All except me had a parent who was a scientist or an engineer. At Berkeley, we had regular meetings of the approximately 20 women graduate students, who probably made up less than 10% of the physics students. At IBM, a group of us women occasionally went to lunch together off-site. One time a male postdoctoral fellow asked if he could come, but we said no because we typically had lunch with him every day. At Davis, I was only the second woman faculty member in physics. We now have 6.1 women FTE, or about 14% of the physics faculty.

Reflections and Recommendations

People change, and you will, too. You will want different things at different times in your life. Remember that you can

73

change careers. Perhaps you will not want to be a scientist all your life. My chemist friend from IBM is now an internationally known quilt, glass, and jewelry artist. Another IBM colleague moved from experimental physics to software development.

I initially did not want an academic job because the lifestyle of my thesis advisor looked unappealing, plus I did not particularly want to fight my way up the tenure ladder. After 11 years in industrial research, however, that same lifestyle was attractive. Unfortunately, as industrial research has become more applied (more development and less academic-style research), it has become harder to go from industry back to academics, although the reverse transition continues to be easier, i.e., from academics to other careers, such as entrepreneur.

Good opportunities can come your way unexpectedly, and you need to make the most of them when they occur. Take the opportunity to learn new things and to become an expert in different areas – scientifically, administratively, and for the soul, even artistically.

Remember that your career is not your whole life. My family is far more important than my career. I have been fortunate to have a supportive husband, who also has a Ph.D. degree in physics, and three marvelous children, two adult sons and a daughter still in high school. We are a family of alpine skiers (the children are all ski racers!). My mother moved to Davis just before our daughter was born. She provided our daughter's early daycare and has continued to help by cooking many dinners and driving our daughter around town. I also do sewing and quilting with computerized sewing and embroidery machines.

If you have a talent for science, nurture it and follow your dreams. Hard work can make them attainable.

JANET CONRAD

Bona Fides

Dr. Janet Conrad is a Professor of Physics at MIT working on neutrino experiments. For nine years she was the co-spokesperson of the MiniBooNE experiment, and she is now working on its successor, MicroBooNE, which will run in 2014 at Fermi National Accelerator Laboratory. She also collaborates on the Double Chooz experiment located at the Chooz reactor. Lately, she is developing new experiments that use high-power cyclotrons to produce neutrinos at high rates. Janet was an undergraduate at Swarthmore College. She received an M.Sc. from Oxford University and a Ph.D. from Harvard. She was a postdoc and then a professor at Columbia University, before moving to MIT. Janet is a fellow of the American Physical Society and, in 2001, received its Maria Goeppert Mayer Award for her leadership in experimental neutrino physics. She has also been a Sloan Fellow and a Guggenheim Fellow.

Always a Detective

I think if my twelve-year-old self came to visit me in my office at MIT, she would like it. She would like the bouquet of green scintillating fibers, the giant copper Frankenstein-like knife-switch on my desk, the antique spectrometer in the corner, and all the physics-related toys and widgets on the shelves. That said, I think she would also look around and demand to know, "Where's the magnifying glass? Where's the fingerprint powder? The 'wanted' posters? And where are you locking up the bad guys?" Because, at twelve, I was sure I would be a detective.

Being a detective was not my first career choice. Even earlier I had wanted to be a scientist: either an astronomer or the chief science officer on a star ship; I could not decide. Once I reached the age where I could start reading Nancy Drew and Sherlock Holmes, however, my whole worldview changed. I was going to solve mysteries!

It seemed the perfect career for me. It was exciting. It required intelligence and dogged investigation, and I was (and am) nothing if not relentless. And as a detective, I could be as different as I wanted to be and still be very cool; after all, few people are weirder and, at the same time, cooler than Nancy and Sherlock.

Well, this career plan did not quite work out. I discovered I don't like blood, and as I learned from TV, that's a problem for detectives. So I shifted back to pursuing my earliest love – science – and I found I did not have to give up the mysteries that motivated me. In fact, if you think about it, this way I could do both: a detective is not always a scientist, but a scientist is always a detective.

The Mystery of the Silent Neutrinos

Now I am a physicist. My job is to pursue the most elusive of the matter particles in the universe: neutrinos. They are quite hard to catch, requiring enormous, intricate, and beautiful detectors. Also, it turns out all of the hours I spent "studying" my mystery novels (instead of doing homework) are coming in very handy. Consider this from the short story "Silver Blaze:"

> Scotland Yard Detective: "Is there any other point to which you would wish to draw my attention?"
>
> Sherlock Holmes: "To the curious incident of the dog in the night-time."

Detective: "The dog did *nothing* in the night-time."

Holmes: "*That* was the curious incident."

Indeed! And sometimes neutrinos do the same curious thing. Sometimes neutrinos go silent and cease to interact in our detectors. I want to solve this mystery – I want to know why.

The existence of neutrinos was first proposed in the 1930s. As a matter of fact, we are awash in neutrinos, though we did not realize it for many years. Among the lightest of the subatomic particles, they pass through the earth in all directions, day and night. They originated in the Big Bang that began the universe, and they come to the earth from exploding stars and, most commonly, from the sun. We can also make our own neutrinos, using reactors and accelerators – most of my work involves these man-made particles. In recent years we have discovered there are actually three distinct neutrinos. They are called electron, muon, and tau flavors, though vanilla, strawberry, and chocolate would be just as good (maybe better).

Observing neutrinos is a lot of fun because it requires giant detectors. I helped to build one called MiniBooNE, for small (Mini) Booster Neutrino Experiment. This detector is a 40-foot-high sphere that contains 800 tons of mineral oil. This is "mini" in the world of neutrino physics, where detectors can be more than 150 feet tall. The detector is filled with photosensors that detect the flashes of light that occur when neutrinos interact with the oil. These sensors, called photomultiplier tubes, work on the principle of the photoelectric effect, first explained by Einstein, where certain metals produce electrons (an electric current) when hit by photons (light). The metals that do this are a lovely amber color. The photomultiplier tubes line the inside of the sphere, which is painted black. Before the detector was closed and filled with oil, I would stand at the bottom and look up – it was like a strange night sky filled with more than a thousand moons. If you would like to build truly beautiful detectors, neutrino physics is for you.

Using many detectors like MiniBooNE, we have discovered that – unlike any other matter particle – neutrinos seem to come and go. Sometimes they interact; sometimes they don't. The effect was first observed with solar neutrinos, but now it has been seen in neutrinos produced by accelerators and at reactors as well. We now think the effect is periodic, and so we call it "neutrino oscillations." Apparently, rather than disappear, neutrinos morph from one kind into another and back again. So the neutrino changes its type from one that can be seen in the detector, to one that cannot be seen. Let's say it starts as a chocolate neutrino. At some point later in time, it may evolve into strawberry. But if your detector only "tastes" chocolate, it seems like there is no neutrino there.

The "smoking gun" to a good detective would be the observation of the other flavor. And, in fact, in the last few years we have also observed neutrino "appearance." We have built detectors that differentiate strawberry from chocolate. We have seen that a pure chocolate beam does evolve into a strawberry one with time.

So, over the last ten years we have developed a credible case for what neutrinos are doing. But why? We need a motive! And this is what interests me now. We know from these results that the Standard Model, our set of laws for particles, is being broken. According to our rulebook, neutrinos do not oscillate. The fact that they do oscillate and seem to go "silent" in our detectors, could be – like the dog in the night – the clue we need!

The Secret of the Standard Model

Why do I think there is something quite different and more fundamental underlying the Standard Model? We have too many elementary particles: at least twelve matter particles,

78

twelve antimatter sisters to these and four force carriers and a Higgs. These particles can also be arranged according to their characteristics, like a periodic table. In fact, when we have holes in the table – when we are missing particles with characteristics that would fit in – we have discovered that we can go out and find them. That was how we realized that one of the neutrinos, the tau neutrino, had to exist, even though we were not able to find it until 1999. If the table of clues is that predictive, then it seems elementary to me that this is not the final theory.

The problem is that I cannot look at the clues we already have and figure out what lies beneath. I need to find more clues. And this is what I am working on now. We have a set of "anomalies" in neutrino physics. These are results of morphing neutrinos that do not fit the model of neutrino oscillations we have developed. I am wondering whether these signals mean that other types of neutrinos exist. These new neutrinos, which are predicted in some theories, would not interact at all; they would be flavorless. On the one hand, if these neutrinos exist, that would be really amazing – it would point the way toward a final theory. On the other hand, these anomalies may be false clues: effects that look strange but have a Standard Model explanation. How do I go after a particle that does not interact? I have to pursue every Standard Model possibility, eliminating each one, one at a time. And once I have done that? Well, recall Holmes' advice in The Sign of the Four: "When you have eliminated the impossible, whatever remains, *however improbable*, must be the truth."

Not So Elementary…

So you can see why my twelve-year-old self would be perplexed. My work is done in big labs, not creepy attics and abandoned mansions. I follow tracks of particles in detectors rather than footprints of criminals through the snow. I am intrigued to find flashes of light rather than dropped handkerchiefs. But that's OK, because there is a lot that hasn't changed from my junior high school dreams. I'm still chasing clues and following trails, and like any good detective, there are moments when I put all the ideas together and can suddenly see a solution. There are moments when I know the truth and no one else has seen it yet.

So, now that I am a professor, what sort of advice would my twelve-year-old self want from me? None, actually. She was always very hardheaded and independent. And that's ok – because, in truth, generic advice is rarely valuable. Each of our lives is very different, and none of the advice in this book may be right for you. Giving advice is not so elementary.

If I must give advice, then I recall what Nancy Drew observed in The Hidden Staircase: to succeed you need "moxie and a good sense of balance." Let's face it: to reach the top of any field – whether it is physics or astronomy, if it is as a science officer on a starship or as a detective – you are going to need courage, skill and confidence. Moxie takes you a long way. And balance? Well, Nancy meant physical balance since she was climbing rickety stairs – that does help if you have to climb around on very big detectors! But another kind of balance is intellectual balance. It is important to know about, and be excited by, more than just your narrow, chosen field. Good ideas come from unexpected places. The time you spend absorbing diverse ideas – including detective work – will really pay off in your physics in the end.

My last piece of advice is to look at life as an adventure. In fact, I am convinced that the adventures that brought me to MIT are just as good as any had by Nancy and Sherlock. I wish you the fun of adventures like these!

ESTHER CONWELL

Bona Fides

Dr. Esther Conwell's career as a working physicist spans many years, from 1942 when she got a B.A. from Brooklyn College, to date. After receiving her B.A. she started graduate work at the University of Rochester. When it became clear by mid 1943 that almost all of the faculty had already departed to do war work or would soon do so, it was decided that she would do a master's thesis with Professor Victor Weisskopf, who would still be at Rochester for a few months, in order to have something to show for a year and a half of graduate work.

Interrupted by World War II and by taking a job teaching at Brooklyn College (because she and her new husband decided to live in New York), she finally got her Ph.D. degree from the University of Chicago in 1948. After five years of teaching at Brooklyn College, she achieved tenure and a year of leave, because the teaching line expected to be hers was reclaimed by its occupant, who had been ill and not expected to return. She spent the year at Bell Labs, working primarily for Bill Shockley. Deciding in that year that research was what she most wanted to do, she did not return to Brooklyn College; she spent the next 20 years doing research and publishing in semiconductor physics, starting at Sylvania, which subsequently became General Telephone and Electronics (G T & E), both then in New York City. During that period she wrote a book, "High Field Transport in Semiconductors," which was published as a volume of the prestigious series, Solid State Physics, edited by Seitz, Turnbull and Ehrenreich. During this period she also gave birth to her son, Lewis Rothberg. Years later he greatly pleased his mother by becoming an outstanding physicist.

After the time with Sylvania-GTE and a brief stint as a visiting Professor at MIT, Esther spent 26 years in the research labs at Xerox in Rochester, doing research and publishing on various topics in solid state physics. In that time she became associated with, eventually becoming a member of the Executive Board of an National Science Foundation Science and Technology Center at the University of Rochester, populated mostly by chemists from the University. Upon retiring from Xerox in 1998 she became a Research Professor in the Chemistry Department at the University of Rochester and still occupies that position. Since then she has been researching and publishing on the transport of excess electrons and holes in DNA.

Among the many honors she has received are the IEEE Edison Medal, membership in the National Academies of Science (1990) and Engineering (1980), the American Academy of Arts and Sciences (1992) and a National Medal of Science (2010).

Early Work

This first section aims to tell the story of my master's thesis and also to illuminate the uncertainties faced by one woman and the importance of fighting for one's own interests.

In late 1942 or early 1943 Professor Weisskopf, who was then at Rochester, visited Purdue, where they were doing early research on semiconductors (at that time Germanium(Ge); Silicon(Si) came later). This research turned out to be a forerunner for the invention of the transistor at Bell Labs not many years later. The Purdue people, principally Professors Lark-Horovitz and Vivian Johnson, told Weisskopf about the results of measurements of current vs. voltage on Ge at Purdue. Around room temperature it was found that resistance decreased with decreasing temperature, as was known to be the case for metals, and was well understood to be due to scattering by the thermal vibrations of the lattice. The puzzle was that at lower temperatures the resistance increased as the temperature was lowered further. I imagine they suggested to him that the low temperature behavior was due to scattering by the impurities (dopants) that were added to the semiconductor to give rise to the conductivity. In any case, Weisskopf suggested that doing a theory for the contribution of the impurities to the resistance was to be my master's thesis. He gave me an idea of how such a theory could be set up. Apart from an occasional interview with him to ask questions, I spent a lot of time worrying about whether I was capable of setting up such a theory and whether I could do it in the few months before he left for Los Alamos. The day before he left I handed him an essentially complete theory for the impurity scattering along the lines he had suggested. I did not hear from him again for a couple of years. He was undoubtedly busy, but I do not think he took me seriously as a physicist. When I met him at MIT many years later he remembered me as having been "his first woman Ph.D. student."

The Purdue people, perhaps having heard from Weisskopf that I had come up with a theory for the impurity scattering, were eager to try it out on their data. After a special appeal from one of their students who had been a classmate of mine at Brooklyn College I sent them the theory. Not long after I was informed that, when combined with the lattice scattering, the temperature dependence of the impurity scattering accounted well for the experimental results. Johnson told me where I could find a theory for the lattice scattering, due to Bethe, and I also could verify that there was a good fit between theory and experiment. There was no question of publication because the Purdue people had gotten the theory, as well as their results, classified as war work. About a couple of years later, when I was working toward my Ph.D. at the University of Chicago, I happened to look at the newly arrived Bulletin of the American Physical Society and found that at the next APS meeting Conwell and Weisskopf were giving a 10 minute paper on the theory of impurity scattering, followed by a 10 minute paper by Lark-Horovitz and Johnson on the successful comparison of theory with experiment for resistance vs. temperature of Ge. Apparently the work had been declassified. I decided to go to the meeting, which was in New York. Lark-Horovitz claimed he had not been able to find my address, and had been planning to give the Conwell-Weisskopf paper himself, but did allow me to present it. Subsequently I wrote a paper on our theoretical work, which Weisskopf approved, and it was published in *The Physical Review*.

Discrimination

My early career provides examples of the flagrant discrimination against women characteristic of physics until ~1960 or 1970. Of course, discrimination continues to this day, but some of what I underwent would be illegal now.

My earliest experience of discrimination came in 1943 when I was still a graduate student. Not much was happening at the University of Rochester that summer, so I decided to try for a summer job in New York, also the home of my boyfriend, later to be husband. On the strength of about a year of graduate work I was hired as an Assistant Engineer by Western Electric, part of the then monopoly Bell telephone system. A couple of weeks later my boss took me upstairs to see his boss. The latter informed me that they had no payroll classification for women as Assistant Engineers. I would have to be demoted to Engineer's Assistant, and it is clear what that did to my salary. It was undoubtedly legal, and since the job was only to last 7 weeks in total, there was nothing to do but continue to work for the 7 weeks. I assume that the point in hiring someone like me was to get an early look at a person they might hire as a permanent employee upon his acquisition of an advanced degree; apparently it only occurred to them belatedly that a woman might apply for the job.

Some years after I started working at Sylvania I decided to try for a job at the more prestigious IBM Research Labs in Yorktown Heights. I was duly interviewed by scientists and management and lunched with a number of the physicists employed there. But finally, some weeks later I was told I would not be hired because IBM had a rule against hiring married women.

When I became pregnant, I informed G T & E that I would be taking a leave for 3 months after the birth and then would return to work. The company, which on the whole had treated me more or less as they did male employees, even giving me some administrative responsibility, handled this by terminating my employment. They rehired me when I returned after the 3 months, but this treatment turned out to cause a big reduction in the pension I ultimately got from the company.

Of course, my experience of discrimination did not end there. But it was more subtle, like job offers I did not get, or not getting invited papers at conferences that I felt I would have gotten if I were a man. I do not doubt that although the flagrant instances I experienced earlier are rare today, other women in physics continue to experience the more subtle forms of discrimination.

Careers in Academe vs. Careers in Industry

My career differs from that of most physicists in that I have spent a majority of my time in industry rather than the academy. This affected the focus of my research; I was encouraged to do physics that had some relevance to the company's products, although this criterion was very liberally interpreted. The relevance requirement generally did not conflict with publishing my research results in *The Physical Review*, for example. I found that the relevance could also be a source of professional satisfaction. The standards set for the research, at least in the best companies, were on a par with those set in the better academic institutions.

Despite the above considerations, most women in physics and related disciplines tend to seek careers in the academy rather than in industry. As a member of one of a number of committees set up to further the opportunities and treatment of women in science and engineering, I ran a conference out of which came the book, "Women Scientists and Engineers Employed in Industry: Why so Few?" The book was published by *National Academy Press.* Happily the numbers have increased somewhat since then, but women are, of course, still quite underrepresented in physics. Among the outcomes of that conference was the realization that many companies had addressed the problems of professional women, including those stemming from pregnancy and child rearing, in a manner far superior to that of the universities (Corning was cited as an outstanding example). For example, at least one company installed a facility that made it possible for women who were breast-feeding to continue. Some big companies set up day care facilities for the children of employees. A large university might be able to do the same, earning the gratitude and improved functioning of male as well as female employees. Adjustments in the tenure clock would also be particularly helpful to women with children.

JILL P. DAHLBURG

Bona Fides

Dr. Jill Potkalitsky Dahlburg has been Superintendent of the Space Science Division (SSD) at the Naval Research Laboratory (NRL) and a member of the U.S. Navy Senior Executive Service since December 2007. In this position she leads conception, planning, and execution of space science research and development programs with instruments to be flown on satellites, sounding rockets and balloons, ground-based facilities, and mathematical models. Dr. Dahlburg served as NRL Senior Scientist for Science Applications from June 2003 to December 2007. Her duties included facilitating/expediting the accomplishments of the scientific missions of organizations within NRL, with emphasis on interdisciplinary areas of opportunity and distributed autonomous systems. From 2001 to mid-2003, Jill left NRL to work for General Atomics as the Director of the Division of Inertial Fusion Technology and Co-Director of the Theory and Computing Center. In 2000, she served as Head of the NRL Tactical Electronic Warfare Division's Distributed Sensor Technology Office, where she co-proposed and was co-principal investigator for the first year of development of the small, expendable, unmanned aerial vehicle Dragon Eye, which saw active duty in Iraq. Jill began her federal career at NRL in 1985, working as a research physicist. As a member of the NRL Nike KrF Laser Program from its inception through 1999, she contributed to laser matter interaction research, implosion and coronal hydrodynamics, and laser beam imprinting. Her work included spearheading the development of the first three-dimensional multi-group radiation transport hydro-code appropriate for laser-plasma modeling. Jill holds a B.A. degree in liberal arts (1978) from St. John's College in

87

Annapolis, and an M.S. degree in physics (1980) and a Ph.D. degree in theoretical plasma physics (1985) from the College of William & Mary. She is Chair of the American Physical Society (APS) Panel on Public Affairs (2011-2012), 2012 Chair-Elect of the APS Mid-Atlantic Section, Chair of the Navy Space Experiments Review Board (2007-), and 2012 Past-Chair of the APS Topical Group on Energy Research and Applications. Her previous professional service includes serving as 2005 Chair of the APS/Division of Plasma Physics, Member of the Lawrence Livermore National Laboratory (LLNL) Defense & Nuclear Technologies Director's Review Committee (2001-2006), and Member of the National Research Council Committee: Quality of the Management of S&E at the Department of Energy (DOE) National Nuclear Security Administration [NNSA] Laboratories (Phase-I, 2011; Phase-II, 2012). Her honors include six NRL Allan Berman Awards for scientific publication excellence and a DOE Appreciation Award presented by DOE Under Secretary for Science Raymond L. Orbach for outstanding service as the Chair of the DOE Advanced Scientific Computing Advisory Committee. Jill is a Fellow of the APS.

Then

My interest in the fantastic secrets that nature offers began when I looked out of our screen door on a summer night, as a very small child, and watched the light pattern cast by the moon on the door's black wire screen. It wasn't round, like the moon, but rectilinear, like the screen's wires! This thoroughly puzzled me, and I pestered everyone I met to find out why this happened, why the moonlight spilled out over the screen in such an unaccountable pattern. None of my family being scientists, they could not really explain, but they tried – particularly my father. Mostly, they just said that I would understand when I got older.

Years after, I devoured books like those by Herbert S. Zim, in particular, the one about insects that recommended looking under backyard rocks with a hand-held magnifying glass. Ugh!

That decided me: the biological sciences were not my destiny. Later, I had an exceptional high school physics teacher, who shared with me the wonders of the physical world, and how it could be understood with mathematics (yes, finally, even the light of the moon through a metal screen). Miss Sharon Fortna, who taught my 11th grade physics class in 1972, conveyed to our class not only science but also the important idea that "If you don't understand something right away, don't despair, just relax and let your brain work on it by itself." This was a new concept for me, and a revelation when I found out that it actually worked.

In 1974, I chose St. John's College in Annapolis for my undergraduate education. The College advertised that year that we would study mathematics from the beginning, and I wanted to understand the whole story of math, from Euclid to now, and why it was so essential to our civilization. The treatment of science (natural philosophy, as it was called there) kept me riveted. I could not master how to explain my interpretations of Baudelaire in a way that swayed anybody over to my view, but a math proof or science experiment? They were right or wrong, facts were facts, and there was no need for rhetoric. I was hooked.

Plasma Physics Research and Development

I left St. John's with my B.A. in 1978, entered the College of William & Mary, and became the student of David C. Montgomery, an inspiring plasma physicist. He introduced me to the technical path of computational plasma physics, and he also enabled my interview with plasma physicists at the Naval Research Laboratory (NRL) after I received my Ph.D. in 1985. I was hired at NRL and spent many happy years focused on increasing fundamental understanding of laboratory, space, and astrophysical plasmas and related magnetohydrodynamic, gas dynamic, hydrodynamic, radiation, and atomic basic physics phenomena. I also applied my knowledge as a plasma physicist and high performance computing simulationist to the design, development, and interpretation of plasma-based experiments and systems in topical areas including: laser matter interactions and inertial confinement fusion, magnetically confined fusion energy, high energy density and warm dense matter laboratory experimentation and comparison with space and other plasmas observations, and directed energy. It was great fun.

Nearly fourteen years after I began work as a research physicist at NRL, I had a discussion with NRL Director of Research, Dr. Timothy Coffey. I confided to him that I was happy with spending my thinking hours on research, but that I was beginning to feel a pull towards something that would allow me to peg my professional contributions more directly, to society and to the Laboratory. He grinned a huge grin (unfathomable to me at the time), and within what seemed to be a matter of hours, had me assigned from my research bench in the NRL Plasma Physics Division to working for Dr. John Montgomery, Superintendent of the Tactical Electronic Warfare Division (TEWD), and partnering with the Head of the TEWD Vehicle Research Section, Richard Foch. John's favorite maxim, "Learn by doing," stood me in wonderfully good stead during the coming months!

Dr. Coffey – who often said that we don't need many science managers, we just need a few really good ones – had with this reassignment to John Montgomery's office placed my feet firmly and irrevocably on the path to Scientific Management (hence the grin: watch what you wish for!). It is this step in my science journey that I want to highlight in this essay. Like Dr. Coffey, I believe that we do not need a lot of science managers, but we do need a few, and I hope that of those who are reading here some will be among that number.

My Research as a Plasma Physicist UAV Developer, or How I Became a Science Manager

Soon after my arrival to TEWD in late 1999, Rich Foch and I collaborated to co-propose an over-the-hill small Unmanned

Aerial Vehicle (UAV) reconnaissance initiative, Dragon Eye. In February 2000, the Office of Naval Research (ONR) committed funding to NRL for the whole of this three-year initiative. Rich and I began joint service as Dragon Eye's Co-Principal Investigators for its first year. We assembled and led an NRL team to develop Dragon Eye and engage warfighter experimentation with our prototypes; we achieved the hosting of an Industry Day within one year after receipt of funding (on February 7, 2001, barely squeaking in under our self-imposed deadline); and, in 2008 Dragon Eye became an exhibit at the National Air & Space Museum as the first of a kind organic UAV that supported the warfighter directly[1].

The Dragon Eye success began with a technically credible White Paper on the topic of affordable, expendable sensor air platforms that was provided by TEWD to the Secretary of the Navy at his request, in April 1999. It continued with an act of entrepreneurship in December 1999, when TEWD researchers proactively arranged to demonstrate the NRL Micro Tactical Expendable [MITE] micro air vehicle in reconnaissance visual-imagery mode at the Marine Corps Warfighting Laboratory [MCWL]. To our delight, during this open-air demonstration flight, the MCWL Commanding Officer personally stopped by and queried about the purchase of ... say, 100 MITEs for the Marines. Over the next few weeks, Rick and I acted promptly to develop a concept for a small, robust over-the-hill reconnaissance nested system of inexpensive UAVs. We targeted the cost for the smaller of these UAVs to approximate that of an affordable mortar round, and we scoped the capability to be fully autopiloted for easy and practical warfighter operation. In early February 2000, Dr. Montgomery arranged for Rick to brief our concept to the MCWL Commanding Officer. On the spot, MCWL requested a proposal from TEWD for a stand-alone version of the smaller

---

[1] http://www.nrl.navy.mil/media/news-releases/2008/nrls-dragon-eye-on-display-at-the-smithsonian-national-air-and-space-museum

of the two UAVs, envisioned as back-packable, few-pound organic reconnaissance asset that – since the year 2000 was the Year of the Dragon – would be christened Dragon Eye. Just one week later, the Dragon Eye full request for funding was committed by ONR and MCWL to NRL, as the Secretary of the Navy's Small UAV Initiative.

Our Dragon Eye proposed timeline and funding were reasonable, but tight. A major risk was the autopilot, which Rick and I estimated would be available within a six-month timeframe from proposal funding award. With a goal of bringing the small Dragon Eye capability to the warfighter as soon as possible, we recommended to proceed with this risk, a decision that turned out to be warranted. To develop our operational prototype on time and within the minimal budget, we were required to hold the entire NRL Dragon Eye multidivisional development team closely accountable for measurable high-quality, timely, and cost-effective output for all aspects of the program. This included technical development, prototype demonstration and delivery, programmatic review, incorporation of warfighter feedback, and an actionable path towards industrial manufacture.

Dragon Eye, subsequently built by industry to NRL design, provided considerable acknowledged in-theatre utility. For example, in a March 2007 Navy news story, <u>Dragon Eye Flies High over 26th MEU, Kenyan Army</u>, Marine "Cpl. Wallenstein said the Dragon Eye '...is invaluable. It gives the troops on the deck an advantage over the enemy.' ... It can be programmed for almost any mission a unit may require"[2]. The Dragon Eye accomplishment confirms that forefront technical development for the warfighter is achievable upon demand within the framework of a world-class government civil service research laboratory. My never-stop, always climbing, whirlwind year as co-PI of the Dragon Eye program convinced me that

---

[2] http://www.navy.mil/submit/display.asp?story_id=28281

the diverse and challenging landscape of science management is truly worth spending my heartbeats.

Do you want to become a successful science or engineering manager, too? The rules are simple. Be positive about your research, your colleagues, and your organization. Be creative and explore new ideas. Take risks. Build beneficial collaborations. Get visibility for your research. Pursue with a grand stubbornness and determination the work that you think is most important. Always tell the truth about your research, and be prompt with your reporting. Give credit where credit is due. Have a vision. Perhaps most essentially, never give up.

Now

I am currently Superintendent of the NRL Space Science Division (SSD), leading a broad-spectrum research, development and experimentation program to study the atmospheres of the sun and the earth, the physics and properties of high energy space environments, and solar activity and its effects on the earth's atmosphere, and to transition these capabilities to operational use. In SSD I share in the amazing opportunity to watch imagery from our instrument hardware on satellites such as *Hinode*, STEREO, SOHO, and Fermi open up the skies and show us a world of wonder – a world that can be discerned and explained by physics and mathematics – and, accordingly, by us. From my earliest professional scientific opportunities as a beginning Naval Research Laboratory federal researcher to my present role as the head of the SSD, I have defined and pursued collaborative world-leading research, and developed and built consensus toward visionary technical solutions for problems of international interest and national importance. It has been a true honor and a great pleasure, and a path tremendously worth taking!

ARATI DASGUPTA

Bona Fides

Dr. Arati Dasgupta received her Bachelor of Science degree in Physics, cum laude, from the University of Maryland in 1973. She went on to earn a master's degree (1976) and doctorate in atomic physics also from the University of Maryland in 1983. After graduation, she worked for a year at Goddard Space Flight Center/NASA in Greenbelt, Maryland and then held a postdoctoral position at the University of Maryland. She joined the Naval Research Laboratory (NRL) in Washington, DC as a research scientist in 1985 and became permanent staff within a year. She remains an international expert in several areas of theoretical atomic and plasma physics. Her numerous awards and professional honors include induction to the Sigma Xi Sigma honor society and Fellowship in the American Physical Society. She has made outstanding and enabling contributions to basic and applied atomic physics with a broad spectrum of activities, and her substantial list of publications includes major contributions to both atomic and plasma physics.

Arati is active in professional and educational outreach efforts at NRL and beyond. She is the current president of the NRL chapter of the Women in Science and Engineering (WISE) network. As a division leader for the NRL mentor program, she routinely mentors high school and college women students as they pursue careers in science and engineering. She is actively involved in nurturing scientific collaborations,

94

with a number of experimental and theoretical Atomic, Molecular and Optical (AMO) groups in the USA, Europe, and India, to help address the AMO needs of the global atomic physics community. She serves as a member on many committees, reviewer for several prestigious journals, and panelist and organizer of numerous workshops, including the Indo-US Science and Technology Bilateral proposals and workshops.

The Early Years

It is a common belief that environmental influences play a critical role in the career path of an individual, especially in the fields of science and mathematics. While this may be true to a large extent, my career path would not have been predicted by the circumstances of my childhood. I was born to a conservative family in Bengal, India. No member of my family pursued a career in the physical sciences, and I was not encouraged to do so either. Whereas the top local schools were taught in English, I attended a mediocre Bengali school that was chosen because of its proximity to my home.

Despite a lack of direct encouragement, I was extremely interested in mathematics and science from an early age. I also became very fond of literature and read volumes of Bengali texts including the writings of Nobel laureate poet Rabindranath Tagore. Although I had a great interest in reading novels and poetry and writing poems and short stories, my real passion was solving mathematics problems. During my pre-teen years, I was often saddened that I did not know enough science or mathematics to solve the challenging problems that I encountered. That motivated me to work harder. I searched for books and other materials in mathematics and science beyond those used in the regular school curriculum. In the 1960s, my only means to locate such texts was through library searches.

During this time, I studied the lives of many famous and successful people. I read biographies of great mathematicians and scientists: Isaac Newton, Albert Einstein, and Marie Curie, to name a few. I also continued to read the writings of literary figures such as Tagore and William Shakespeare. Their stories gave me inspiration and taught me how to pursue my studies with determination. They gave me insight on how to find joy rather than discouragement in the struggles that necessarily arise when solving difficult problems. By reading their stories, I started to dream that one day I too could make a contribution to science.

I was very fortunate that although I attended an average secondary school, I learned physics from two extraordinary teachers who made the subject fun and exciting. Around the same time, my future husband began tutoring me in physics. The inspiration I received from all of them had everything to do with why I chose to be a physicist and had an indelible effect on my eventual professional career.

Juggling Two Lives

A very recent feature article ("Why women still cannot have it all," Atlantic magazine, June 2012) has sparked many lively debates and discussions about the nature of success for women. Although the author's career is in law rather than science, she deals with the same conundrums of success that a woman scientist does. We have come a long way in reducing the gender gap, and women today have greater status than ever before, but it still remains that what is considered "standard" for men is "having it all" for women. Although women are coming close to shattering the glass ceiling and are outpacing men in many arenas, especially education, many women still struggle to find balance between family and career and live with constant feelings of guilt for not doing full justice to either family or profession.

Due to personal circumstances, I married very early and

shortly thereafter came to the United States with only a high school degree. I could not speak English fluently, and I was terrified about going to college and majoring in physics. But I was determined to do precisely that and whatever else was needed to fulfill my ambitions. Because of discrepancies between the Indian and American secondary school systems, I did not have enough credits to enroll directly in a university. I therefore enrolled as a part-time student at a junior college, taking English, physics, history, and mathematics. I also worked part time to pay for my tuition, as my husband was only a graduate student at the time. After a year, with excellent grades in all my courses, I was admitted to the University of Maryland with a merit scholarship.

Much to my surprise, I was one of only two female physics majors in a class of about 30-35 students in my sophomore year. By my junior year, the other woman had left physics for engineering. I was certainly in a unique situation – a married immigrant woman in a crowd of young boys who could hardly relate to me or what I was thinking.

At the end of my sophomore year, I had a son named Sandip, and I took leave from school the following semester. I returned to school and graduated – with honors in physics – within two years. I prevailed primarily because of my strong determination combined with the enormous care and support I received from my husband. Despite the discouragement that I received from some others who told me to focus solely on raising my family, I remained resolute in pursuing my career goals. I was always bold, and I spoke my mind, especially in the face of unjust or unfair circumstances, and was undaunted by opposition or unfavorable consequences. Some may have construed this attitude as brassy, but it strengthened my self-confidence and later helped me gain attention and professional respect in many situations.

After graduating from college, I decided to attend graduate

school at the University of Maryland to get a master's degree and Ph.D. in physics. My decision to attend UMD rather than explore options at higher-ranked universities was primarily based on my family situation. My husband's job was near UMD, and I did not consider living separately from my husband and son. It was very important to me to raise Sandip myself and dedicate whatever time and energy was needed in doing that.

After I received my master's degree in atomic physics, I was blessed with another son, Samit. I was fully determined to continue my Ph.D. work while taking care of my two sons. Due to their age difference, Sandip and Samit required different kinds of attention. This made balancing my life between family and research more challenging. During this time, I also had teaching and research fellowships that involved additional responsibilities. I went to the university for just a few hours during the day while taking care of Samit. I would also take care of Sandip when he came home from school. I studied later at night after they went to bed. I was very committed to spending quality time with my kids, and I spent as much time as I possibly could to nurture their physical, emotional and intellectual health. My husband also did everything possible to take care of our kids and also to support my career. After a few years I earned my Ph.D. in atomic physics. As before, I did not apply for jobs outside of Maryland, primarily because of my family.

After a couple of years as a postdoc, I was offered a research position at the Naval Research Laboratory in Washington, DC. NRL is a premier federal government laboratory with a high level of research activities in many fields of science and engineering. After just one year, I was offered a permanent federal position as a research physicist at the laboratory. However, I still had to worry about balancing time between work and my two boys, who were growing up faster than I could keep up. I used to spend as much time looking for interesting and challenging books for them as I did looking for books and research journals for myself!

My first project at NRL concerned the "Star Wars" nuclear-power weapons program. I was responsible for conducting detailed and complex atomic physics calculations to explain the discrepancies between theoretical predictions and experimental gain in X-ray lasers. I published many journal articles on this topic. I worked on several interdisciplinary atomic and plasma physics research projects that spanned basic atomic, plasma and astrophysical research as well as fusion, lasers, lighting, and other gaseous electronics applications. I have been fortunate to receive recognition for my work, which was considered to have significant impact on major large national experimental facilities.

While pursuing these projects, I was still very heavily involved in the education of my two sons, who participated in math, science and debate teams and competitions. My children's achievements augmented the satisfaction that came from my own research and accomplishments.

Reflections and Recommendations

It is quite unfortunate that despite advancing towards equality for women during recent decades, women are still underrepresented in scientific research, especially in mathematics, physics, and engineering. Women in science face challenges not only in obtaining science-related jobs but also in receiving other associated benefits such as salaries, promotions, and access to resources. The conundrum we face is that if we are not very vocal, we may be branded as insufficiently committed, but if we do speak up, then we may be labeled as too aggressive. The goal of enjoying an equitable work environment where women can thrive and perform their best and receive equal rewards for their work has unfortunately not been achieved. To make matters worse, there is often a veil of suspicion, as some colleagues believe that women enjoy an unfair advantage because of their minority status in the field.

Being in physics, most of my colleagues are men. I have learned to defend my position and not get derailed by any negative or inappropriate comments. By working hard and focusing on my work I have accomplished quite a bit, and I have received accolades that have helped me garner my colleagues' respect.

One of my outreach efforts is to encourage young women to pursue careers in science, for example by giving lectures at local high schools. I love meeting young women with an interest in science, and I often encourage them to apply for a summer position in my laboratory. It can take just one experience to ignite a passion that leads to a career. Often the young women I meet do not feel that a career in physics is realistic, but I can offer myself as a counterpoint. I came to the United States as a teenager with only a high school degree, and I did not speak English fluently at the time. Through my passion for studying physics and mathematics, commitment to hard work, and support from my family, I was able to achieve my goals. The young women I meet have excellent chances to become successful scientists if they make this their goal as well.

In order to empower women in science, efforts must be exerted at the grass-roots level. Concretely, this involves creating opportunities in science education for girls even before secondary school. Though the education and employment environment in the last decades for women in science has witnessed a positive transformation, and women have made enormous progress in the areas of science and technology, barriers still remain. Many women face gender discrimination in their families and at their work places, even in developed countries. It needs to be emphasized that girls and boys are born with equal talent for mathematics and science. If a girl wants to study science and is willing to work, nothing can stop her from being successful.

SARAH M. DEMERS

Bona Fides

Dr. Sarah M. Demers is a particle physicist and an Assistant Professor in the Physics Department at Yale University. As a member of the International ATLAS Collaboration, she uses data from the Large Hadron Collider at CERN in Geneva, Switzerland in her research on elementary particles and the forces that govern their interactions. Sarah received a Bachelor's degree in Physics from Harvard University in 1999 and her Ph.D. from the University of Rochester in 2005. She is on the executive committee of the users organization that represents the 1000+ American physicists whose research is based at CERN. In 2011 she received an Early Career Award from the Department of Energy for her work at ATLAS. Sarah is engaged in trying to increase the participation of women and minorities in science. She is also passionate about communicating her research and increasing the math and science literacy of young people and the general public. Recently she has been engaged in work that puts artistic and scientific disciplines in dialogue. She teaches "Physics of Dance" and "Physics of Music" and co-leads an interdisciplinary research working group with professional dancer-choreographer Emily Coates at Yale.

"Have you been in the women's bathroom recently?" The group of male physicists interviewing me for a postdoctoral research position looked almost as mortified as I became, hearing myself ask the question. I had complemented them for having a woman-friendly environment a few minutes earlier. They wanted details, and I was flailing. The hole already dug, I jumped. "There are free tampons! Little signals like that add up to making women feel expected." The awkward silence that is death to an interview was broken mercifully, sort of, by an emeritus Professor sitting next to me who chimed in with, "Yes! And those bathrooms should have condoms!"

Starting Out: A Skeptic

I grew up believing that women could do everything that men could do. I assumed that everybody thought this, and I was easily annoyed by conversations around feminism. As an undergraduate physics major at Harvard I avoided the "women in science" group, in spite of their offerings of free food. I learned how to use the machine shop, minimize my exposure to epoxy, and not drop things in my research with Professor Melissa Franklin. We were gluing wires to thousands of gold mylar sheets that would improve the uniformity of the electric field in the tracking chamber of the CDF experiment at Fermilab's TeVatron. The majesty of the detector we were upgrading pulled me into the world of particle physics soon after the discovery of a fundamental particle, the top quark, in the mid-nineties.

I didn't become alarmed by the lack of women in physics until graduate school. I was, reluctantly, interviewed by a sociologist investigating the gender gap. Fermilab, where approximately 10% of the associated scientists were women, was a good case study. We went out for Chinese food with an accomplished fellow graduate student who was one of the brightest physicists I knew. She confessed between bites, "I'm not actually very smart. I can only keep up by working harder than everyone else." My brain lurched to reconcile the brilliant scientist in front of me with "not actually very smart." I scrambled to think of something to say to the researcher, because I had been about to use the exact same words. What I ended up with was, "There may be a problem."

At the time I was just referring to my friend's opinion of

herself. If she wasn't actually very smart I would need to claim the "dumb as a doornail" spot on the spectrum. I began to see a pattern of promising young women selling themselves short even before the practice of science had a chance to do more productive pounding. My sense of my own potential may have come crashing down, but my desire for justice was fully intact and activated.

Confidence and Gender Bias

I became the graduate student adviser of a women in science group for undergraduates at the University of Rochester. In our meetings I encouraged them to not feel dumb, though I did. I suggested that they ask questions in class, though I hadn't. And I told them to have the confidence that I was lacking. This was not one of my stronger plans, though we did eat a lot of chocolate. I kept hunting for what might be giving gender this power that it didn't deserve.

We have recent evidence, thanks to research by Yale's Jo Handelsman and her colleagues, that science faculty members, both men and women, have a bias against female students. A resume from a John results in higher rankings and more pay than the identical resume from a Jennifer, who is evaluated as more likeable yet less competent.

Certainly, not all women lack confidence and not all men are full of it. But even a small trend in that direction would be amplified by the sub-conscious bias that so many of us are learning we have.

What can we do to create space for women in science within our own minds? I ask this question on the heels of the discovery of another particle, this time likely the Higgs boson. How much more progress can we make by opening up scientific pursuits to people who could contribute effectively but who are not participating? How much better could we all be if we put less energy into self-doubt and more into progress? As you think about yourself as a current or future scientist, I hope that you will take your potential impact on whatever field ignites your passions very seriously.

I have not personally compiled the full list of items that should be available in university bathrooms or a general set of best-practices to improve the climate, though this kind of information is available. What I can do is advocate for changes that remind us that we expect women to be equal contributors to science. Oh, and it turned out that in spite of – or because of, or unrelated to – the fact that I brought up tampons in the interview, they did offer me the job.

MILDRED
DRESSELHAUS

Bona Fides

Dr. Mildred (Millie) Dresselhaus had a humble start in New York City. She was trained in the liberal arts at Hunter High School and Hunter College, then, after a year at Cambridge University and another year at Harvard, she studied physics at the University of Chicago. After a two-year postdoc at Cornell, she started her independent career as a research staff member at Lincoln Laboratory. She worked on high magnetic field studies in the semimetals bismuth and graphite and continued to use high magnetic fields intensively until 1990, when the National Magnet Laboratory moved to Florida. In 1968 she was promoted to full Professor of Electrical Engineering at MIT, and in 1985 to Institute Professor. She entered the nanoworld of research in 1973, and since then she has worked broadly on many topics in condensed matter physics. She was President of the American Physical Society in 1984, President of the American Association for the Advancement of Science in 1997, Treasurer of the National Academy of Sciences from 1992-1996, and Director of the Office of Science of the Department of Energy in the Clinton administration 2000-2001. She has been active in training a steady stream of Ph.D. students and in mentoring both undergraduate and graduate students and also postdocs, especially women, and including many visiting students who stimulated entry into new research areas in condensed matter.

Early Years

There are relatively few women from my generation in physics. When I finished my Ph.D. thesis in physics in 1958 at the University of Chicago, the number of physicists overall was quite small, and the percentage of women getting Ph.D. degrees was even smaller (2%). So I start this article with my early years before completing my Ph.D. I grew up in New York City in a dangerous, multiracial, low-income neighborhood where the schools emphasized discipline more than learning. The only people I knew who had a high school or higher education were the school teachers and the family doctor. Working in my favor was my only sibling, a brother who was three years older than me; he was a child prodigy both academically and musically. When I was four years old and he was seven, he was offered a full music scholarship for studying violin and associated musical studies. When the teacher at the music school saw that I could sing all his pieces, they soon offered me a junior scholarship for violin lessons, allowing me to use his hand-me-down instruments and music. It was through music school that I learned about the importance of education, self-reliance, and high standards for achievement. (I made good enough progress on violin so that I could keep my music scholarship through most of high school). It was also through the music school that I learned about the existence of Hunter High School, the only special public high school available for girls at the time. Entrance to this special high school in New York City was by examination, and I learned that it was possible to obtain copies of previous examinations by request. Perusal of these examinations showed me that I was a long way from passing the examination, but New York City had good libraries and kind librarians who were helpful in locating appropriate books for self-study, which is what I did. The experiences in preparing myself for these examinations put me in a highly favorable position in gaining a deep understanding of many basic concepts, in developing self-discipline, motivation, and purpose that I would not have otherwise experienced at a young age.

My entry to Hunter High in the tenth grade was the first turning point in my life. There I found, for the first time, a strong academic program and a student/faculty environment with a focus on academic advancement and excellence. My first year at Hunter High largely involved catching up on all the subjects that were not tested on the entrance examination. In my second year I suffered from a medical setback by coming down with whooping cough, which was highly contagious, so I was not allowed to attend school until this long illness was over. However, keeping up with school work by self-study, this setback was surmounted.

As graduation approached, I had to make a decision about what to do next. The guidance counselor at Hunter High laid it out clearly for me: my three options were school teaching, secretarial work, or nursing. My preference was strongly for school teaching. Hunter College, right next door to Hunter High, offered an excellent program for teacher training, which seemed like a great opportunity for me. While in high school, I had had a variety of private tutoring jobs with students struggling in a variety of academic subjects. I enjoyed this work and was pretty good at motivating the youngsters to study, focus, and gain understanding.

Thus I became a student at Hunter College with the intention of becoming a school teacher. Because I was well-prepared for college through my excellent high school training, I could handle a big course overload at Hunter College, taking both the required courses as well as courses that I was interested in. I was also still able to keep up my tutoring jobs to earn money (and pick up teaching experience). I learned many things at Hunter College that were very important for my later career, including time management, that women could make it in academics, and how to take advantage of opportunities when they appeared. Some teachers took a special interest in me and made good suggestions for self-study. In particular, I was much influenced by Rosalyn Yalow who was the teacher in my sophomore modern physics class at Hunter College. It was she who in no uncertain terms told me to focus on the study of physics because it was a field about to take off. She thought that I would find many opportunities if I went to a good graduate school and worked diligently. Rosalyn Yalow wrote letters of recommendation for me to the best graduate schools. Although I lacked depth when I completed my B.A. degree at Hunter College, I had breadth and a high level of motivation. Toward the end of my undergraduate years I saw a poster notice about a Fulbright Fellowship program and decided to apply for it because it provided money to study in a foreign country. Because of my own very limited financial resources, I had not gone further from my New York apartment than one could go with a rented bicycle. Therefore foreign study seemed very attractive and glamorous, and it was for this reason that I applied for a Fulbright Fellowship.

My year as a Fulbright Fellow at Cambridge was a second turning point in my career. At Cambridge I attended lectures on many subjects and met very talented people from all over the world. At Cambridge, I also learned that I had a lot of hard work ahead to catch up with students trained at the top academic institutions. After a year at Cambridge and another year at Harvard, I wound up at the University of Chicago in the fall of 1953 as a physics graduate student.

My Physics Training

It was at the University of Chicago that I learned physics. In my first year in the physics graduate program I took Professor Enrico Fermi's course on quantum mechanics, which taught me how to think like a physicist. Fermi was a fabulous teacher and a great inspiration. His idea was that students learn physics by working problems. Every lecture was inspiring and enjoyable and ended with a seemingly simple problem. But actually solving the problem turned out to be less simple and required deep thought. Understanding of physics at the

University of Chicago was tested by rather challenging examinations, one taken at the end of the first year and another after the second year, with only half of the students gaining passing grades on each. I prepared for these examinations by working problems on past examinations, and this endeavor gave me a broad understanding of physics. Fermi believed that students should have a broad competence in physics so that they could pursue thesis work in any physics field where opportunity presents itself. This turned out to be useful at many points in my own career, and I am thankful for the broad physics education I received at the University of Chicago as a graduate student. The fact that graduate students had to come up with their own thesis topic as well as the method for how to carry out the research itself was not easy, but it was valuable training that I appreciated in my later career.

On Being a Woman in Physics

My experience as a woman in physics is somewhat unusual. As I mentioned above, I had a difficult childhood. Coming from an immigrant family with little means, going to poor schools until tenth grade, and having a father with a serious mental illness made being female a very minor factor for me. Even though I was not allowed entry to the science-related special high schools because they were for boys only, I didn't know enough about discrimination to be personally affected, and it was also too minor an issue to bother me. I did gain entry to a terrific high school that served me well. I did attend an undergraduate college that prepared me well enough to obtain a start on a scientific career. I even gained confidence about being a woman in physics because Hunter College, though being officially an all-women's college at the time, did open its doors to World War II veterans when I was an undergraduate. These veterans tended to have a lot of practical experience with radar and other technical things they learned from military service. They therefore taught me a lot about laboratory work, which they knew from their wartime experiences. In the classroom the women students generally performed better on the problem sets and the examinations, so I did not get the message that women could not do physics when I was an undergraduate.

At the University of Cambridge where I spent a year as a Fulbright fellow, the number of women physics students was small in number, but I was at a Women's College, Newnham College, where studying physics was unusual but well respected. At Harvard University, I was a student at Radcliffe College, which again was an all-women's college. Again, studying physics was unusual, but there were two women graduate students in my year, and we both did well in our coursework and exams. My first negative experience came at the University of Chicago. There was only one faculty advisor in condensed matter physics available to take on graduate students; he did not believe women should study physics, and he told me so. However, in my first year, I had met Enrico Fermi and got to know him and his family very well; they were very encouraging to me. Unfortunately, Fermi died after my first year, but I was already moving through the examination process by that time.

Fortunately, I met Professor Clyde Hutchinson, a physical chemistry professor. He invited me to join his Saturday lunch club. He taught me a lot about microwaves, which was close to the topic I chose for my Ph.D. thesis, the study of the microwave properties of superconductors in a magnetic field. I also had a female friend, a chemistry graduate student, De Lyle Eastwood, whom I mentored for years, and we supported each other a great deal during graduate school. In fact, even though there were few women students, and we were in different years of the program, we all supported each other and helped overcome the prejudice against us. We had a role model, Maria Mayer, who was a very successful scientist at the University of Chicago, and I got to know her well. Even though she received a Nobel Prize for the work she was doing when I knew her, she still could not get a faculty appointment at the University of Chicago (though she was allowed to work and have graduate students). For me that was good enough, and I did not have high expectations for myself—the opportunity to practice as a physicist was quite sufficient.

I was treated very well at the University of Chicago; I had a fellowship or other departmental support continuously. Some people appreciated me while others just tolerated me. For those who told me to "get lost," I did so. I can, however, end this stage of my career on a happy note, because my official faculty advisor, who wanted me to get lost, got back to me 10 or 15 years later with a big apology, saying that he was wrong and I was right about entering a physics career. In this way, the story ended happily—we became good friends, and he became a supporter of women in physics.

After my Ph.D. at the University of Chicago I spent two years as an NSF postdoctoral fellow at Cornell University continuing study of the same topic. I chose to do my postdoctoral work at Cornell because I married Gene Dresselhaus in 1958, while completing my Ph.D. thesis, and Gene was a young physics faculty member at Cornell. During the second year of my postdoc, I was told by the department head that my performance had been below expectations and that I would not be allowed to continue there as a researcher, even as a volunteer. In hindsight I would agree with the department head that he made the correct decision for all concerned.

Therefore Gene and I sought employment elsewhere, and we received very attractive offers from IBM and the MIT Lincoln Laboratory. Preferring an academic environment, we chose to join the MIT Lincoln Laboratory in June 1960.

My Independent Career

My independent career started with magnetoreflection experiments in graphite and bismuth using high magnetic fields. I also took advantage of teaching opportunities when MIT faculty, who also were working at the Magnet Laboratory, were traveling. Thus when the Dean of Engineering started looking for somebody who could teach physics to engineers I became a candidate. At about the same time the Provost's office was looking for someone to fill the newly created Abby Rockefeller Mauzé professorship for a woman visiting professor, and so I was also nominated for that opportunity. It was in this way that I became a visiting full professor in the MIT Electrical Engineering department, and within a few months it was converted to a tenured appointment in 1968.

Working in a stimulating environment, with excellent graduate students, a steady stream of highly motivated and talented visitors, and postdocs worldwide, led to a very exciting and stimulating career. For the past 52 years, I've worked at the cutting edge of various areas of condensed matter physics, nanoscience, and applied physics. I have continued learning physics from and with my students, enjoying all aspects including research, teaching, student mentoring, and service to my profession and my university. Working at the cutting edge I have had opportunities to meet my counterparts worldwide and to start collaborations with them. Through my membership in the National Academy of Engineering and the National Academy of Sciences I get involved in science policy, both in the U.S. and worldwide. I have found these committee activities and the people involved in these activities stimulating and inspiring. Thus I wake up every morning looking forward to a day of new happenings and scientific adventures with all my coauthors and collaborators worldwide.

Reflections and Recommendations

As I look back at my adult life, it couldn't have been more rewarding and enjoyable. I love my daily scientific life, my family—husband, four children, and my children's families including five grandchildren—and am in good health at 82 years of age. I enjoy being a researcher, a co-author, a mentor, and a research collaborator. I also feel good about giving something back to society as a payment for all the free education and opportunities that came to me over the years. I strongly recommend this rewarding path through life to others interested in science and service to society.

LUCY FORTSON

Bona Fides

Dr. Lucy Fortson is an Associate Professor of Physics in the School of Physics and Astronomy at the University of Minnesota. A member of the VERITAS and CTA very-high-energy gamma-ray astronomy collaborations, she studies Active Galactic Nuclei (AGN) using multi-wavelength observations to determine the source of gamma-ray emission from AGN and the evolution of the AGN host galaxies. Lucy is also deeply committed to improving the science literacy of all Americans through her role on the Executive Committee of the Citizen Science Alliance and the Zooniverse project (www.zooniverse.org). With projects such as Galaxy Zoo, the Zooniverse provides opportunities for volunteer citizens to contribute to discovery research by using their pattern matching skills to perform simple data analysis tasks and to become more deeply engaged in the science research through social networking and simple data processing tools. She was recently the Vice President for Research at the Adler Planetarium in Chicago where she held a joint research position at the University of Chicago. Lucy graduated with a B.A. in Physics and Astronomy from Smith College and received her Ph.D. from UCLA in high energy physics. She has served on numerous local and national committees including the National Academy of Sciences Astronomy 2010 Decadal Survey as co-chair of the Education and Public Outreach committee, the Astrophysics Science Subcommittee and the Human Capital Committee of the NASA Advisory Council (NAC), the Mathematical and Physical Sciences Directorate Advisory Committee (MPSAC) for the National Science Foundation, and the Education and

114

Public Outreach Review Committee for the National Optical Astronomy Observatory. She is a member of the American Astronomical Society and the American Physical Society. In 2010, she was awarded NASA's Exceptional Public Service Award.

A Stellar Inspiration

My story begins with the skies of Seattle in the early 1970's. I was about ten years old. We lived in a house that overlooked Lake Washington in just such a way that when I was lying in my bed at night, I could see the stars rising up over the lake in the east (when it wasn't raining). It was pre-Microsoft, so the skies were darker back then. I invented names for the patterns of the stars I saw – the Big Tent surrounded by circus animals, for example. Later, I was informed that the correct name for this constellation was Orion. But when I was ten, the stars were mine.

I grew up in a neighborhood full of boys, so I stayed inside and read a lot (when my brother hadn't cajoled me into being one of his soldiers in the ever-raging Capture the Flag campaigns played in the local woods). My brother and I also shared an interest in catching and caging various Northwest species of newt, snake, lizard – much to our mother's dismay as she would always find the remnants of the poor creatures that escaped into the dark recesses of closets. We loved to tickle our brains with conversations about infinity and what was there (if anything) before the beginning of the universe. I read everything about natural science I could get my hands on. I remember getting the list of new Scholastic books I could order at school: 6th grade versions of everything from new theories of plate tectonics to how hurricanes worked. It felt like Christmas when the order finally arrived. When I was fourteen, my father, who was a physicist at the University of Washington, gave me the book "The First Three Minutes" by Steven Weinberg. I was floored. How could we know so much about how the universe formed? Never mind the stars, I wanted to own the universe! I wanted to understand the whole thing.

At Garfield High School, I was one of three girls in my physics class, and the most outspoken. Our teacher liked to tease me in lab, "If Lucy can do it, anyone can." That made me feel pretty awkward. I had a better biology teacher, Ms. Massey. She knew I loved astronomy and put me in touch with a program at the Pacific Science Center called Motivated Astronomy Student Seminars (MASS). I got to meet astronauts who were part of the new Space Shuttle program, build telescopes, and go on trips to see eclipses. Once, for my language arts book report, I went to the University of Washington Aeronautics Library, lied on the cool floor between the stacks, read the latest technical brief on SkyLab, and wrote up my report on that. I can't remember the teacher's reaction, so it must have been OK. By then, I knew I wanted to be an astronaut. I was inspired as a young child, watching the events of the Apollo program unfold. After scoring very high on the Armed-Services Vocational Aptitude Battery (ASVAB), I was heavily recruited by the Air Force. I wanted to fly jets (I knew astronauts flew jets). But back then, they didn't let women fly jets. So I told them "no" and decided to go to Smith College instead – one of the nation's most prestigious women's colleges.

Smith College: Finding Myself

I think a lot of people were confused by my decision to go to Smith. I got along well with boys (as a tomboy, myself) and was very uneasy around girls. I could understand boys; they had a direct, though sometimes harsh, language. Girls talked in tangents, lacing their seemingly passive prattle about appearance with aggressive sideways cuts. I most definitely did not fit in with the cheerleaders, but I was also not happy to be relegated to "mousy, quiet nerd." I wanted to be popular, so I tried to wear make-up, do my hair just so, and shop for the exact right clothes. I wouldn't go out with the boy-nerds for fear of being branded. In the end, my best friends were two boys who were from "intellectual" families; they didn't question my brains or my femininity. So it was a strange decision, even for me, to decide to go to a women's college. I remember thinking that if I wanted to study physics, I didn't want to have to put up with the male arrogance exhibited in my high school physics classes. So off I went to the east coast.

I majored in both physics and astronomy, but it was very difficult for me at first. The fact is, I did not know how to study effectively. My high school, it seemed, had been too easy for me, and all of a sudden I was with hundreds of other students who were just as smart, or smarter (!) than me, AND many of them had learned the discipline of studying to survive their high schools. I nearly flunked out my first semester. I was put on the Dean's "watch" list. My competitive nature rose to the challenge and by my senior year, I was on the Dean's list for graduating with a very high GPA in my major fields and writing a senior thesis in nuclear physics theory.

During my years studying physics and astronomy at Smith, I finally began to feel a sense of belonging. There were a few other women at Smith who shared my passion for learning everything they could about how the world worked. These were women who spoke my language: yes, nerdy, but proud of it. Together we fought the social stereotypes that wanted to strip us of our femininity and deny us of our thoughts about boyfriends and fashion. The messaging was all around us – in movies, advertising, broken hearts: Strong, smart women met sticky ends – or more mildly, if you wanted to wear a "power suit" you would scare away any chance of marriage and family. These were battles fought even on the "inside" of the physics department as one of the older physics faculty told me "you are too romantic to be a physicist" (later, much later, when I was a post-doc at the University of Chicago and back visiting Smith, this same professor told me he had been "very wrong" about me! Hah!). But mostly, my physics classmates and I had a good time studying together, grappling with Differential

Equations and Electricity and Magnetism, filling the blackboards in our study room with wrong turns and right solutions as we learned physics along with the pleasures of true friendship.

Lest I leave a wrong impression, Smith had several really nurturing faculty in both physics and astronomy who were excellent in teaching and their research, most notably for me, Suzan Edwards and Margaret Pfabe. Dr. Pfabe in particular was my senior thesis advisor and was so encouraging. She signed me up to give a talk on my thesis at the 1984 APS meeting and worked with me night and day to get me ready for this debut into the professional world. I received several very positive comments after my talk including one from a man who made a point of saying he was so happy to see a young woman giving such a good talk and how important it was for me to keep going because I was obviously very talented. Interestingly, several of the faculty at Smith were women, and I know that their existence helped solidify my own thoughts about what was possible as a career with a physics degree.

UCLA: A Peek into High-Energy Physics

Even so, I was not completely sure that I wanted to continue on in graduate school, and I took a research assistant position with the Space Telescope Science Institute in Baltimore. There, I realized that with a B.A. in physics and astronomy the career path was limited, and I wanted to learn more physics! Off I went to graduate school in physics at the University of California, Los Angeles. I was one of four women admitted in fall 1985 with a class of approximately sixty incoming graduate students! Sixty is a large number – the apparent strategy of the department was to admit large classes, since they needed teaching assistants for the large service courses, but then reduce class size by eliminating half after the year-end qualifying exams. Thus into the meat-grinder I went, slogging it out through Goldstein and Jackson and all the other "bibles" of physics. I was learning so much, but would it be enough?

Thankfully, yes. But so many of my classmates didn't pass that first year. It was brutal, but I look back and see that it was really by studying for the qualifiers that I learned physics.

My first big project as a graduate student was doing detector development for a space-borne gamma-ray observatory. I worked with Professor David Cline who had recently arrived at UCLA. Unfortunately, it quickly became obvious that this project was not going to provide me with a thesis. As it happens, Dr. Cline was also on the UA1 high energy particle physics experiment at CERN and needed a graduate student to move there and take on responsibility for reconditioning some of the detector elements. This was a very exciting opportunity: this was the very same experiment that had just detected the W and Z bosons (the mediators of the weak nuclear force) and provided the spokesperson, Carlo Rubbia, with a Nobel Prize in Physics. I joined the marching army of graduate students tasked with keeping the data flowing from this giant device that peered at the tiniest of particles released from collisions at the highest energies human-kind could make. I graduated with my Ph.D. in high-energy physics in 1991 with a rather uninspiring thesis. I realized I didn't like working on large 400-person collaborations. I also realized how much I missed astronomy and armed with detailed knowledge of the smallest-scales, set out to work my way back to owning the universe.

A Return to Astronomy: Particle Astrophysics

My timing couldn't have been better. The notion of particle-astrophysics was just coming to fruition; by understanding particle interactions at very high energies, we could begin to understand the earliest stages of the universe. Furthermore, the window of high-energy astrophysics was opening up with new gamma-ray detectors, which probed the sources of extremely energetic phenomena: from the supernova death throes of stars in our own galaxy to the centers of active galaxies whose jets spew material to distances of millions of light-years. With my experimental high-energy physics tool-kit I was in the right place at the right time to take on some of these great new challenges. I received a post-doctoral position at the University of Chicago to work on one of the biggest gamma-ray detectors ever built: the Chicago Air Shower Array (CASA).

These post-doctoral years were truly amazing, and I felt all my years of graduate school training came together. I was also fortunate to be in an excellent collaboration that encouraged initiative and leadership. By the end of my second year, the main work of the collaboration was finished and many of the senior people were moving on to other projects. I seized this opportunity to utilize the CASA infrastructure and (with the help of a great graduate student working with me) designed, built, installed, and operated 144 "lightcones" that we used to detect the faint blue flashes of light that accompany the charged particles in airshowers. This Cherenkov light could tell us about the types of nuclei that were hitting our atmosphere as cosmic rays and, more importantly, could give us hints as to whether or not they were created in supernova. Our measurement, published in 2000, the year my son was born, still stands today as the best Cherenkov measurement of the composition of cosmic rays in the energy region where supernovae are important. What a sense of accomplishment! What pride! All my hard work had been worth it – after many horrible hours filled with self-doubt, questioning whether I was good enough to pass my exams and whether I could make it past my own insecurities – I was finally able to contribute something significant to the fantastic body of knowledge that humans are painstakingly building up.

One might think that I would take such an accomplishment, get a good faculty position somewhere, and ride off into a golden tenured sunset. Indeed, I received several very good faculty job offers. But my story has never been a simple story. One day, sitting on the lawn at the University of Chicago with my husband, trying to sort out where to go next, Dr. Evalyn Gates, head of the Astronomy Department at the Adler Planetarium and a friend of mine, came along. As it turned out, the Planetarium was trying something new: hiring scientists who would conduct research at the same time as carrying out public outreach programs. I took the bait. I had always felt a keen desire to "give back" to the public and bring them along on the amazing journey that is scientific discovery. After a few years, I became chair of the department and subsequently, Vice President for Research. I picked up where Evalyn left off and grew the Research Division into a strong, vibrant group with research in practically all wavebands from the solar system, to cosmology, the history of astronomy, and education research. Along the way, I learned a thing or two about designing exhibits and sky shows, public lectures and educational programming.

Blazing a Trail in Science Education

Once again, I found myself in the right place at the right time. The national science funding agencies, responding to a crisis in science education, were calling for scientists and educators to take up this challenge of educating the next generation. A new vital career path was opening up: the science education and outreach professional. I thus had the opportunity to serve on some of the highest-level committees at the National Science Foundation and NASA to advise these agencies on how to best commit their resources to improve science literacy across our nation. I became particularly excited about the possibilities that the Internet provided in building a community of "distributed researchers" of all education backgrounds. I began to work with colleagues on developing tools to allow high school students access to data from the Sloan Digital Sky Survey and worked to get programs started at the Adler that engaged the public in real research. The culmination of all of this work was when I teamed up with Chris Lintott from Oxford University and the scientists who had started Galaxy Zoo – the most successful online citizen science project at the time – to start the Citizen Science Alliance. This group runs the Zooniverse.org set of citizen science projects with over 750,000 people around the world contributing to projects that range from classifying galaxies to images of lions in the Serengeti.

The Long Way Around to Becoming Faculty in Physics

I am living proof that in this day and age, one can leave academia for 13 years as I did, and in the end come back. In 2010 I took a faculty position in the School of Physics and Astronomy at the University of Minnesota to solve a two-body problem: the "my-spouse-and-I-need-two-jobs-in-one-city-please" difficulty of modern academic couples. I am happily teaching physics, continuing my research in gamma ray astronomy, and working to promote citizen science and human computation with the Zooniverse. This is probably the most important lesson I think I can provide for young people entering into the field today: it is possible to follow your heart and do what you think is important work, and your career is what you make it. For some, the road may be seemingly easy. For the rest of us, use your passion and perseverance to struggle against your self-doubt. It is a hard road, but it is definitely worth it.

ELSA GARMIRE

Bona Fides

Dr. Elsa Garmire graduated from Radcliffe College (Harvard University) in 1961 with an A.B. in Physics. She received her Ph.D. in Physics from the Massachusetts Institute of Technology (MIT) in 1965 and was a research fellow at Caltech for 9 years. She went to University of Southern California (USC), Center for Laser Studies, in 1975 as a member of the research staff. In 1981 she became a full professor with tenure at USC, the first woman out of the entire 120-member engineering school faculty. In 1989 she was elected to the National Academy of Engineering and in 1992 became the William Hogue Professor at USC. In 1995, Elsa became Dean of Engineering at Dartmouth, the first woman dean of any research engineering school, and she was elected a member of the American Academy of Arts and Sciences. She has since returned to research in lasers and optics and holds the title of Sydney E. Junkins Professor. She has authored over 200 journal papers and holds 10 patents. Her recent interests include Technological Literacy for all Americans and International Policy in Communications and Information Technology. She spent 2007-2008 as a Jefferson Science Fellow with the U.S. State Department.

Becoming a Physicist

My father was a chemical engineer and the first person in his

123

family to go to college; my mother was a homemaker and violin teacher. As the second of three girls, I took on the role of the boy my father never had. When as a 6th grader I was given a pamphlet entitled *Careers in Science*, I focused on the last page, which showed a "research scientist" in a white coat (a man, of course). At that moment I knew this was the career for me. My choice of physics was determined by my belief that physics was the most "pure" of the sciences. We explain how the world works. Besides, biology in those days (before DNA) was pure memorization, and a summer job in a chemistry lab taught me that smelly chemicals held no appeal.

As a woman determined to go to Harvard, I applied to Radcliffe College. We were admitted to a separate college and lived on a separate campus, but all our classes were together, as were our exams, which were graded on the same scale. We were not allowed into the Harvard undergraduate library but rather had our own library at Radcliffe (where we could bring our boyfriends!). With 300 women at Radcliffe and 1100 men at Harvard per class, we knew merely by the numbers that Radcliffe's students were, on average, much better than Harvard students. This encouraged self-confidence unavailable in typical co-ed schools.

I majored in physics (with two other women) and took advantage of every opportunity to match my educational experience to that of Harvard students. I arranged for a post-doc to come over to Radcliffe and teach us about relativity. I approached a favorite professor to ask if I could help with research. This led to my designing and carrying out an experiment in optical pumping. I felt so important, reading original research papers in the graduate student library! Today we know that research–oriented projects inspire women to stick with science when the going gets tough; I am a living example of this fact.

I signed up to grade papers, which brought me close to Roy Glauber, a Nobel prize-winning theoretical physicist. Equally valuable was Nobelist Ed Purcell's magnificent teaching of quantum mechanics. My senior year was enlivened by acting as Teaching Assistant (TA) for a freshman physics laboratory, using undergraduates as TAs, which had never been done before at Harvard. I was chosen as one of this select cohort and the only woman. As a freshman, my first laboratory class had been lonely; no one wanted to partner with the only woman. Laboratory classes (as well as problem assignments and exams) were challenging because they assumed more familiarity with mechanical systems than I had. I had never fixed cars, nor played with model trains. Nonetheless, I enjoyed my first laboratory experiment immensely: measuring the thermal conductivity of copper – a simple experiment that taught error analysis. That I loved this lab, while most of the boys found it boring, may have been an indication of how much I would love experimental physics. It still sends shivers up my spine when my students' experiments agree with our predictions.

When it was time to apply to graduate school, I did not feel very confident. My grades were half A's and half B's (Harvard had no minuses or pluses). I needed to stay in the Boston area because I was marrying a graduate student at MIT (whom I'd met at Harvard). My hero, Ed Purcell, encouraged me to apply to both MIT and Harvard, though I thought I wasn't good enough for either. Purcell knew better. I had sat in the first row and asked questions because his classes inspired me immensely. After class I invariably jumped up with a string of additional questions. Shyness hadn't kept me from quenching my thirst for physics knowledge. Unknowingly, I was demonstrating my capacity as a serious student – the first step toward success. This Nobel prizewinner was responsible for my excellent education and future career. He was arguably my first mentor.

Graduate School

To my surprise I was admitted to both Harvard and MIT with

full funding. It may sound silly, but I decided to go to MIT because I was tired of the Harvard library, where I had done all my studying. Put another way, I realized that new vistas would await me at MIT. I recommend to everyone that they go to a different graduate school than undergraduate school, for the same reasons.

Walking down the halls of MIT on the first day was terrifying – the main hall is about a half-mile long, with everyone scurrying about on important scientific business. I felt like I was back in high school. However, I fit well into classes – even if I was the only girl. The only silly thing is that I dressed up to go to class with a girdle and high heels! My only role models for dress were secretaries. I wanted to be lady-like and not "manly." I even slip-covered my slide rule case with red velvet and sequins! It helped that I was married, so I was freed from the incessant worrying about finding a husband that dogged undergraduates at Radcliffe. I quickly degenerated to jeans.

My classes inspired me, but I cannot remember at all how well I did. It no longer mattered. I was learning interesting and important material that has served me well throughout my career. Understanding became more important than grades.

I was not so happy in research, however. I had been placed with a female research faculty member working on X-ray crystallography of minerals. I found the research topic boring, and there were no other students with which to interact. I felt like an outcast among the male students. By Christmas-time I seriously thought about dropping out. However, my husband pushed me to go back and seek another opportunity. Fortunately, MIT's new provost, Charles Townes, was looking for a graduate student. Little did I know that he would win the Nobel Prize as the originator of Quantum Electronics – demonstrating the world's first maser and introducing the laser. We had a good meeting, and I found myself working in the Spectroscopy Laboratory at MIT. Real hands-on experimentation was just what I wanted.

Looking back, my days in the Spectroscopy Laboratory seem like magic (which I hope to write about on another occasion). I was joined by another student, Ray Chiao; we worked well together as our skills were complementary. The challenge I gave myself was to explain a 10% discrepancy between experiment and theory in laser-induced stimulated Raman scattering. Sometimes I think the whole point of graduate school is to prove you can persevere on a problem until you solve it, no matter how difficult it is. This is what it seemed like to me. My pig-headedness caused me to spend many hours repeating experiments with slight changes, hopefully working towards agreement. During this time I wasn't exactly dejected, but I was stymied. I reduced my time in the lab and veered into political activity, as a presidential election was approaching. I'd like to say my brilliance solved the problem, but the answer came from an experimental fluke. Perhaps my brilliance was in recognizing the result when I found it; as Pasteur said, "Chance favors the prepared mind."

As my Ph.D. research was simmering along in its third year, I developed a biological urge for a child. My husband was now an assistant professor at MIT, I was 25, and it seemed like the time. I was an "old lady" by those days' standards. I was able to manage the timing to simultaneously finish my Ph.D. and have my baby. (Townes and I had decided I could stop and write up my results when I could no longer reach the optical setup!) I was finishing the last chapter of my dissertation when the baby arrived. I proof-read my thesis in the hospital and defended my Ph.D. two weeks later. Without this constraint, I could easily have spent another year or two exploring my newfound results. It's always nice to have a time-deadline to finish your thesis.

Post-Doctoral Life

After my Ph.D., I went home to raise the baby, having been brought up with attitudes of the 1950's. While I loved my baby, I found being at home full-time was hard work and not particularly rewarding. I missed the excitement of research. Prof. Townes was able to find me a post-doc position, I found a day-care opening, and I spent the following year juggling the baby's needs and full-time work.

After a year, my husband and I moved to Caltech, where he had a faculty position. I took a post-doc in Caltech's engineering school, working with Amnon Yariv. At that time Caltech had no women students, nor faculty. However, they did accept women post-docs. I was the first woman in electrical engineering; my social life revolved around my husband's position.

I was lucky to be able to balance my work and my home-life. I arranged to work part-time, from 10 AM to 4 PM (75% time). We rented a house in the same block as my work, so no time was lost commuting. I had a housekeeper who took care of the baby, cleaned house, and cooked dinner. A second child came 2½ years after the first, over the weekend, and I went back to work Monday morning (just for a short time – I think I was trying to prove a point!).

Life wasn't as simple as it sounds, however; I was torn between trying to carry out original research and being a good mother. Few middle-class women worked at that time, and I felt guilty because I didn't need the money and worked only for enjoyment. (It never occurred to me that my research was a contribution to the world of science – I didn't have that much self-confidence.) While I loved my children, I will admit that they bored me when I was with them full-time. True to that era, my husband didn't do any of the childcare – never changed a diaper! He tolerated my working but rarely discussed my research.

I continued in this position for eight years, working with graduate students, but not leading anything. Having worked under Townes, and now with Yariv, I had seen what top quality research was, and I had doubts that I was creative enough to do it myself. I lost confidence in my ability to create research ideas. Seeking an outlet, my intellectual interests got swept up in the "Art and Technology" movement that was rapidly burgeoning in Los Angeles – but that is a story for another time.

I did, in fact, eventually come up with my own research idea, which I took to Yariv; he supported my writing a proposal for independent funding. The first National Science Foundation (NSF) funding in Integrated Optics and a number of papers resulted from my idea. Next, I designed and built a new laboratory for epitaxial growth of semiconductor lasers and was again involved in exciting research.

At this point my husband and I decided to go on sabbatical for a year. He was a visiting professor at Cambridge University, and I commuted to the ITT Standard Telecommunication Laboratories at Harlow, Essex, outside London three days a week to carry out the first Integrated Optics experiments in England. I also wrote a classic reference chapter on Integrated Optics in semiconductors.

On My Own

After this sabbatical year my world fell apart – a story I tell here merely to show how to get back up on your feet after disaster strikes. My marriage and my position at Caltech ended. It hit me like a ton of bricks! I became ill, retiring to the sofa, unable to get up for several weeks: I had no job and no husband; I was a single parent with two children aged 6 and 8. I even feared ending up on welfare! It's hard to believe how little self-confidence women brought up in the fifties had about their ability to survive in the world alone. There was nothing in my upbringing to offer me confidence, no role models that I could turn to. I had been surrounded by male researchers and female home-makers and secretaries.

By seeking help from others, I eventually located five

"saviors." First, John Marburger, chairman of the Physics Department at University of Southern California, offered me a part-time job in the Center for Laser Studies (limited research money was available). Second, Arnold Silver at Aerospace Corporation also wanted to hire me. I agreed to work part-time at each place, which I did for a number of years, comparing life in a government laboratory with life at a university (my time at ITT had given me experience in an industrial laboratory).

My third "savior" was Judith Osmer, who worked for me at Aerospace. She was politically active in the Los Angeles women's movement of the 1970's, and the first lesbian with whom I spent time. She inspired me to join her in the women's movement. I learned that many of the difficulties I had experienced were the result of gender bias and not my own "fault." I gained a great deal of self-confidence from the "consciousness-raising" group that I formed as part of the women's movement. Six young professional women met weekly and compared our experiences and learned that becoming aware of previously unconscious biases is more than half the battle.

The fourth "savior" was an engineer who would become my husband, Bob Russell, whom I met in 1975. My private life took a turn for the better, and my scientific life flourished as I balanced these two employers with a happy home life. I eventually applied for a regular faculty position at USC. It took an extra year for their decision because of my lack of teaching experience – I had to go on tranquillizers. In 1981 it was finally over! I received an appointment as tenured full professor becoming the first woman tenured faculty in engineering at USC (out of 120 total faculty members). I had been named fellow of IEEE the year before and fellow of The Optical Society of America that year. (My fifth "savior" was Jarus Quinn, director of The Optical Society of America, who had gotten me extensively involved in professional societies – see below.)

Life as a Full-Fledged Faculty Member

I began the life of a typical professor, negotiating for research space, graduate students, and funding, all while trying to continuously generate new ideas. My contacts from Aerospace held me in good stead: I continued there as a consultant. Northrop Corporation provided another valuable consulting experience. Close collaboration with researchers from Jet Propulsion Laboratories, Hughes Research Laboratories, and General Dynamics were also important to generating ideas and funding.

During this phase of my career I built up a sizeable research group. I taught a course taken by the entering class of graduate students, which allowed me to evaluate their abilities early. With broad interests, I opened my doors to many students and pursued a variety of technical directions. Some students were cream-of-the-crop, some were brilliant but for various reasons had difficulties, a number were women, some were black; only a few were "ordinary." I focused on students from under-represented groups, obtaining the first NSF Research Experience for Undergraduates (REU) grant that focused on minorities. I promoted underrepresented women in science and engineering nationally. Ultimately 31 Ph.D. students have completed theses under me, as well as 13 M.S. students. In addition, I've mentored 35 post-docs and visiting scientists. Of these, 10 were women and four were black.

My first independent funding came through the Dean who introduced me to an Air Force Program Director and praised my abilities. I wouldn't call him a "savior," but this act was extremely valuable. I learned then how important it is for minorities to be actively supported by those above them. I was principally on my own because there wasn't another woman in my department from the time I was appointed in 1981 until 2009.

I left USC in 1995, after having been named the William

Hogue Professor in 1992 (the first woman with a named professorship in engineering), to take up a position as Dean of Engineering at Dartmouth, after which I returned to research and teaching at Dartmouth. (That's another story.) This transition into administration was made possible by the visibility I gained through active participation in technical societies.

The Role of Professional Societies

I learned the importance of professional meetings as a graduate student. I was fascinated with meeting authors of papers I had read, and they always seemed happy to talk with me. Later I realized how inter-personal skills are crucial to your career: you need to be known.

My fifth "savior," Jarus Quinn, Director of the Optical Society of America (OSA), provided me with a great deal of self-confidence. He asked me to join OSA's Finance Committee as a volunteer. I had never even seen a spreadsheet before! But the Treasurer knew what he was doing, so I could relax and learn. Technical society volunteerism provides opportunities to learn aspects of management without stress.

In 1977 I was asked to organize a technical workshop in Washington. I had been talking to an Air Force program director about funding my work and knew he'd be impressed that I was running a workshop. All went well, and as a direct outcome of the respect that I garnered at that meeting, I received two Air Force grants totaling $532,000 (in 1978 dollars), worth $1.8 million dollars today! My volunteerism, encouraged by Jarus, set the stage for my career.

Slowly I was chosen on committees with more and more responsibility. Eventually I was nominated and then elected to the governing board of OSA, and eventually to the Presidency. I moved up in leadership at technical meetings, as well, eventually co-chairing an OSA Topical Meeting. Similar opportunities were available in other organizations: IEEE and the American Physical Society.

The National Science Foundation likewise provided leadership opportunities through reviewing grants, running review panels, and serving on the Engineering Advisory Council. I joined the Air Force Scientific Advisory Board and committees of the National Research Council. Various universities including Harvard, MIT, and Australian National University, asked me to serve on advisory or review boards.

These leadership roles placed me in the public eye, and I was elected to the National Academy of Engineering in 1989. With this status, along with my experience in OSA (as well as my position as Director of the Center for Laser Studies at USC), I was approached and asked to apply to Dartmouth to become Dean of Engineering. The school's focus on balancing research with high quality teaching and its interdisciplinary approach synchronized with my ideas, and moving to academic administration sounded interesting. So, my husband and I moved across the country and started on a new adventure, in large part because of my involvement with technical societies.

Reflections

My story shows that many factors play into career success: dreams, acculturation, competition, stick-to-it-iveness, luck, overcoming bias, etc. While at one level I knew since 6th grade that I wanted to be a scientist and dreamt of replacing Marie Curie, at another level, as a product of the 1950's, I wanted to be the world's most perfect mom. I found it challenging to reconcile these two. My innate competitiveness (as a second daughter) and my ability to stick to it when the going got tough (which I attribute to years of piano training) kept me focused. I had good luck at the beginning, but I reached out for it and worked to the best of my ability. I was able to obtain the highest quality education available and whenever I had a choice, I sought excellence over all else.

I believe these factors play into the life of every successful career woman. Professionals have come to understand that expertise takes 10,000 hours to develop. It will take deep, inward dreams to get through 10,000 hours of problem-solving when you begin as a novice. It helps to be competitive, at least with yourself, yet to keep in mind that the culture in which you find yourself will invariably affect your success in competition. You may not make it to the top of the pyramid, but it's fun to be well up there, where the view is broad and international. This climb may require deviations from initial expectations. It requires the ability and willingness to hang on when the going gets tough. Understanding the role of luck helps you accept what the world throws at you – both the good and the bad. You don't do it all yourself. If you're lucky, life will be a positive experience and you'll be glad that you made the climb. For the record, my two daughters are happily married, with their own children. While neither discovered an aptitude for science or engineering, both have been fulfilled by balancing family with their own life pattern (at present one works part-time for a non-profit and the other is a full-time homemaker). My passion for them has balanced my passion for physics and engineering. Along with my spouse, they are the light of my life.

JARITA C. HOLBROOK

Bona Fides

Dr. Jarita C. Holbrook is from Southern California and graduated from the California Institute of Technology with a Bachelor's degree in Physics. She went on to earn a Masters of Science degree in Astronomy from San Diego State University and a doctorate in Astronomy and Astrophysics from University of California, Santa Cruz; she was the first African American to graduate with these advanced degrees from these universities. Her postdoc years included the Max Planck Institute for the History of Science in Berlin and an NSF Minority Postdoctoral Fellowship, which she took to UCLA's Center for the Cultural Studies of Science, Technology, and Medicine. Jarita's first faculty position was as a cultural astronomer in an applied anthropology unit at the University of Arizona – a unit which folded in 2009 leading to her current affiliation as a researcher in Women's Studies at UCLA. Jarita is the current Chair of the Historical Astronomy Division of the American Astronomical Society. She is the first African American to lead a division of the American Astronomical Society in its entire history.

From Physics to Filmmaking

So I Was Good At Math

Though I was born in Hawaii, I was raised in Southern

135

California. My parents were educated professionals with Bachelor of Science degrees: my father's was in science education and my mother's was in nursing. My father later went on to earn Masters degrees in subjects related to social work. My father was on active duty in the Air Force when I was born, stationed in the Pacific and Asia. I was raised with my two sisters: Shawna, older, and Shannon, younger. None of us, nor any of my half-siblings, have followed so-called traditional career paths.

In elementary school I was always good at math. If not the top student I was near the top, and I was a teacher pleaser – that is, I performed well because I liked my teachers. There is great diversity in Southern California, and I enjoyed being surrounded by a diverse student body and diverse teachers. As I moved into the magnet program in middle school, though the general student body was diverse, my classes were not. For the rest of my education through my Ph.D., I was either the only African American, the only woman, or one of just two or three of each in my classes. My friends and colleagues find it hard to believe knowing me now, but I was relatively shy and quiet, especially in my classes, through the time I graduated from Caltech.

I have been asked hundreds of times how I became interested in astrophysics. When you live in Los Angeles and attend magnet school you are exposed to scientists and hear about their careers. Therefore, through interactions with my classmates and teachers, I learned that it was possible to have a paid career as an astrophysicist; so, that is what I decided to do. I think my interest in astronomy was not unique – I find that most kids are interested in the stars and planets at some point in their lives.

Breaking All The Rules and Taking Risks

In middle school I was chosen to attend a special lecture at the Natural History Museum in Exhibition Park given by Dr.

George Fischbeck, a popular weatherman in Los Angeles. Dr. Fischbeck was the first person I heard discuss the realities of being a scientist: how it involves taking risks and how most great discoveries are stumbled upon or due to mistakes. He gave a humorous account of how Pyrex, Benadryl, and Teflon were mistakes found while scientists were trying to create something else. Though I had already decided to become an astrophysicist before going to his lecture, he made science sound exciting, random, and fun, which reinforced my decision.

Being an undergraduate at Caltech taught me that rules exist, but if you can make a good argument, those rules can be changed or ignored if needed. For example: before Caltech I had been a student at California State University, Los Angeles, and North Carolina State University. I was still in middle school when I started attending college (during the summers). My fellow middle school students and I were the exception to the rule of when and who is supposed to attend college. As a result, before setting foot on Caltech's campus I had already taken college level geology, astronomy, and physics, along with high school advanced placement classes in math and the sciences. I learned at Caltech that anything was possible if one just asked; this allowed me to become involved in great research experiences, provided funding to go on cool trips, and let me swap out some physics classes for astronomy ones.

Upon graduating from Caltech I was pretty burned out, and I wanted to travel and see the world. I joined the Peace Corps, but I didn't get an assignment for a year, so I started my Masters at San Diego State University. After I got my assignment to teach high school in Fiji, I went on leave of absence from my Masters for three years. I learned that this was possible from my future advisor, Ted Daub. Each year I had to reapply for leave, but each year they allowed it. At the end of my Peace Corps assignment when all the other volunteers were in angst about what to do next, I knew that I was going to go back and finish my Masters degree.

After completing my Masters I still wanted to travel and see the world, so I went to West Africa where my sister Shawna was a Peace Corps volunteer. After five months of adventure, a job offer from Sally Heap at Goddard Space Flight Center brought me back to the USA. I was accepted into the doctoral program at UCSC and managed to do a fair amount of international traveling (Mexico, Singapore, and Indonesia) during the four years it took me to get my doctorate.

Since getting my doctorate I have made two risky career moves: the first was turning down a tenure-track faculty position in order to switch from being an astrophysicist to being a cultural astronomer. Fortunately, within a year I had two postdoc offers to perform cultural astronomy research. The second was refusing to take a non-tenure track position at the University of Arizona when my department, the Bureau of Applied Research in Anthropology (BARA), folded in 2009. Instead, I have been a researcher-at-large working on NSF funded projects, writing, and making films.

Marriage, Research, Writing, and Making Films

While getting my doctorate at UCSC, I met my future husband, Romeel Davé, and we married after graduation. Despite taking postdocs far from each other we had our first child, and then we worked hard to get double faculty offers. We received offers from the University of Arizona; however, they did not have a history of science department, meaning the best fit for me was with BARA (because I used anthropological techniques for my data collection). Nonetheless, I chose to compromise because being part of a dual career couple requires sacrifices. In the long run, though my junior faculty years were very productive, and I had begun to build a cultural astronomy program at the University of Arizona, the compromise did not pay off, mainly due to my department being merged with another. I resigned from my position in 2010.

My research career as an astrophysicist focused on cosmic gas and dust. At Caltech and during my Masters I studied planetary nebulae looking at the composition, temperature, and structure of the hot glowing gas. At NASA, I studied star formation regions at the center of spiral galaxies using Hubble Space Telescope images. I studied star formation regions in our galaxy using infrared arrays for my dissertation. My career has spanned stellar death to stellar birth, hot gas to cool gas and dust, optical to the infrared. As a cultural astronomer, I studied modern methods of celestial navigation, indigenous African astronomy, and now I'm studying astronomy culture and diversity in South Africa and the United States. I have three books "African Cultural Astronomy" (published), "Following the Stars" (under review), and "The National Astrophysics and Space Science Programme (NASSP)–Creating the Next Generation Of Astrophysicists In South Africa" (just completed).

At the time of my resignation I had already completed my first documentary film "Hubble's Diverse Universe" in 2009. I had wanted to create a film about the Hubble Space Telescope and what it meant to minority astrophysicists. Being from Los Angeles has its benefits, and I knew several filmmakers that I could ask to help me with the project. I was able to secure NASA Education and Public Outreach money to do the film. 2009 was the United Nations International Year of Astronomy (IYA), and I had a leadership role in the activities organized in the United States. I was appointed head of the USA Cultural Astronomy and Storytelling Group. Our group was interested in creating new films that told stories about the sky and having a film festival. However, at the end of the day, my film was the only one that already had funding and thus it was the only one made and shown during IYA2009 from our group. In 2010, after a screening at MIT, I partnered with Enectali Figueroa-Feliciano to make a documentary film about his Micro-X rocket project. Again, we secured NASA funds and started filming in 2011. While still working on Micro-X, I have started filming "Black Sun," a documentary about the 2012 solar eclipses and two African American solar physicists. My filmmaking centers on the lives and science of minority astrophysicists. I am already planning my next film beyond these two current projects.

Discrimination, Prejudice, and Racism

I think that many young people are attracted to science because they like the idea of people who are logical and rational and by extension believe that the world of science is a meritocracy. Women and underrepresented groups in the sciences can often pinpoint when they discovered that the sciences were not a utopia free from discrimination, prejudice, sexism, and racism. Still, many go to great lengths to explain away the bad behavior and actions of their colleagues in order to maintain their illusions to some extent, which in many fosters an underlying discontent. I consider it imperative that women and underrepresented groups be made fully aware that we are all still in the United States and subject to the historical and ever present evils that are part of American culture: do not expect a meritocratic utopia of rationalism.

In moving from physics and astrophysics to the social sciences I found an academic culture that valued diversity not because it was the right thing to do but because of the unique points of view, scientific approaches, and insights diverse scientists provide. I encounter less essentializing situations where people expect me to present THE AFRICAN AMERICAN OPINION rather than just my own opinion. My fellow social scientists can still be patronizing and condescending, but few are dismissive in comparison to my experiences with physicists and astrophysicists. Only a woman or minority can tell you about having been shut down or shut out of discussions because everyone insists on talking over you or around you because their assumption is that you couldn't possibly have anything of substance to add to the discussion. With up-and-coming scientists such actions can be harmful, discouraging, and depressing; however, over time such behaviors can diminish as individuals become more well-known. For myself, I have a keen interest in leadership and have taken on several leadership roles in the astronomy, physics, and cultural astronomy communities. Being a leader and being an expert in several arenas has made me recognizable even in circles where I least expect it.

Being the Change

So the question remains: How do we change physics and astrophysics to make it more welcoming to women and minorities? Since I am now an outsider I play a very unique role in promoting change: my films promote change, my research promotes change, and my books promote change. After many years of being a cultural astronomer, I was surprised to find that my biggest support and the majority of readers come from the astrophysics community. They are hungry for cultural astronomy research on diverse cultures, they love hearing the good and bad stories told by minority astrophysicists in my films, and they invite me to do research on their groups to flush out things that can be improved. I have their trust. I mentor many graduate students who are working on degrees in physics or astrophysics, and I help them strategize on how best to advance their careers. Perhaps most importantly, I talk about them in positive terms to other scientists. Women and minority astrophysicists need an agent/promoter/handler, and I am not afraid to take on that role with the students that I mentor. Don't even think about asking about "quality" and "qualified": anyone who gets into a Ph.D. physics or astronomy program is clearly qualified! The rest is good or bad projects, good or bad advisors, publishing, time, and knowing about the right opportunities. However, a hostile environment can make all of this far more difficult, and thus mentoring is so important in giving students extra support. I invite everyone in this volume to rise to this challenge of becoming both mentors and promoters of their students. Also, I invite everyone to insist upon a zero-tolerance policy for sexual harassment, sexism, and racism in astronomy and physics. Indeed, most universities have such policies in place; we just have to make sure that it is enforced in physics and astronomy, too.

CAROLINA C. ILIE

Bona Fides

Dr. Carolina C. Ilie earned her Ph.D. in Physics and Astronomy from University of Nebraska at Lincoln, one M.Sc. from the Ohio State University, and another M.Sc. from the University of Bucharest, Romania. She joined the Department of Physics at SUNY Oswego in 2008. She is the recipient of the Othmer Fellowship for Exceptional Scholars, American Physical Society Award – Division of Chemical Physics, and Gordon Research Conference Young Investigator Competition Award. She works on National Science Foundation (NSF) ADVANCE grants in the committee on best practices. She is on the senior personnel committee on a $600,000 NSF-SSTEM grant. She serves on the Committee of Women in Material Science and Engineering – Materials Research Society and on the Committee of Women in Physics, American Association of Physics Teachers. Carolina was granted an early promotion to Associate Professor in 2012, and in March 2013 she was awarded the Provost's Award for Mentoring in Scholarly and Creative Activity. She lives in Oswego, NY with her spouse, also a physicist, and their two sons.

My School Years

I grew up in a wonderful environment in Bucharest, Romania. I learned how to read at three years old and how to play chess at four. At six my father prepared me for school (which I started at seven, as was customary at that time in Romania) by filling a math notebook with addition and subtraction problems. My father enjoyed physics, my mother gave me a taste for books, and both of my parents loved mathematics. I absolutely loved math from the first grade.

My parents expected me to perform very well in school. I had the attention of my parents even through the devastation brought to our family by the death of my brother, in a car accident when he was 11; I was 5. Yet in spite of it all, the love and support I received from my parents allowed me to successfully start school and perform at my best. The only challenge I had when I started school was writing: I naturally wrote with my left hand, but in Romania at that time, most elementary school teachers were making an effort not to allow children to write with their left hand. Therefore, I learned how to write with my right hand, and I became ambidextrous.

It wasn't all just studies, of course: I enjoyed roller-skating and riding my bike, like most other kids. There were long summer vacations filled with books, math, and long mountain and seaside vacations.

In Romania, sciences were introduced very early (biology and geography in 5th grade, physics in 6th, and chemistry in 7th), and they were taught continuously until 12th grade. All subjects were mandatory. There was an equal distribution of girls and boys, and everybody was expected to perform well at all subjects. My physics teacher, Mrs. Liliana Pascu, was a very calm, intelligent, and talented instructor. Just two months into the academic term, I came to a realization that her class would shape my life well beyond that year: I told my parents that I would do physics. My mom immediately bought me an extracurricular physics problems manual, which I enjoyed.

The next thing I needed to understand was how to prepare for a career in physics. I did not have what is called a "role model," but I had exceptional teachers for all subjects, including math, physics, and chemistry. Each teacher had at least a Bachelor of Science degree in the subject they taught, and a few of them had master's degrees and even Ph.D.'s. All of my teachers were very passionate about their subjects. I needed a good plan, so I talked to an older cousin who was already accepted (after a very competitive exam) to a top math and physics high school. The high school was located at Physics Platform – Magurele, near Bucharest, Romania's capital. This platform, established in 1956, was the place of work for the Romanian scientific research elite. The Institute of Atomic Physics (IFA), the University of Bucharest – Physics Department, and the "Horia Hulubei" Math and Physics High School are located here, as well as a small reactor facility. Horia Hulubei was the first Director of the Physics Platform and a physicist known for his significant contributions to the development of X-ray spectroscopy. Thus at the age of 11 I knew exactly what high school and university I wanted to attend. On the down side of all of this, I learned about the tough entrance exams; nevertheless, I knew this was my path.

In seventh grade and after a very tough competition, I qualified to compete at the national Olympiad in Romanian language and literature. My Romanian language and literature teacher, a wonderful, realist, and determined woman, congratulated my parents and said that I can do whatever I want: physics, math, or literature. It was my choice, and she would encourage me regardless of my decision. My German math professor, Rainer, said the same but added that physics is a lot of work. I was already aware of this but was determined that nothing could stop me or change my mind.

All went as planned. I successfully passed the necessary high school entrance exam, which covered math (algebra, 2D and 3D geometry) and Romanian (literature, grammar and composition), then the 10th grade exam (very competitive, for both math and physics), and finally the 12th grade end-of-high-school exam (Romanian, math, and physics). After high school, I passed three days of written exams covering math (algebra and analysis), Physics I (mechanics, electricity and magnetism), and Physics II (optics, solid state, atomic and nuclear physics, intro quantum mechanics). Of course, behind these successful exams were countless hours of work, weekends spent with the math and physics books, and tens of thousands of practice problems, yet I can say I spent all those moments in pure joy. I absolutely love, even now, solving math problems (even though I do not teach math anymore) and calculating radicals up to 100, with two decimals, in my head. I am very grateful for all the knowledge I received from the professors I worked with in high school, especially Dr. Marcel Tena and Prof. Petre (math), Prof. Marilena Ruse (physics), and Dr. Gheorghe Zaharia (university math). I am very grateful to my parents for organizing extracurricular math and physics sessions with the best professors in Bucharest and for their overall support.

During university years I learned that I loved both research and teaching. Some of my colleagues were preparing to take their Toefl, GRE and GRE-Subject exams in order to continue their studies in the USA. At that point I had not considered such an option. Instead, I became an assistant professor at one of the top high school colleges, and I took very competitive exams for tenure (300 physicists competed for 7 openings, and I placed third). After 5 years of teaching I thought – now what? I needed new challenges.

My Graduate Years in the USA

My most cherished memories from graduate school were from my classes at Ohio State University, where I had wonderful professors (for example Dr. Ralf Bundschuh, who taught thermodynamics and statistics). However, I most enjoyed my years at the University of Nebraska at Lincoln, where I was surrounded by the most dedicated professors and the best working atmosphere I have ever encountered. Coming from OSU, I was well-received at UNL by all faculty and staff. I am very grateful to Prof. Bernard Doudin, who was my first adviser. He was very encouraging, and the group's atmosphere was very relaxed but highly productive. Another member of Prof. Doudin's group eventually became my spouse.

146

Prof. Doudin left shortly after my arrival for the University Louis Pasteur, in Strasbourg, France. I then joined Prof. Peter Dowben's group, and he eventually became my thesis adviser. Prof. Dowben taught all his graduate students how to work hard, get results efficiently, focus, and construct a future. Even though our group worked very hard, we also had a social life; in the five years I was a part of the Dowben group, there were seven marriages and six babies were born (including my first son). Still, we managed to produce a significant number of papers. Women comprised 50% of the group, which is remarkable for graduate school. Each of us graduated very well prepared, and everybody in the group had at least eight papers and two job offers. If I now make a difference in my students' lives I owe credit to my adviser because he made a difference in mine! How fortunate I was to be part of such a dynamic and productive group!

I started looking for a job in my last year as a graduate student: I received ten phone interviews, four on-site interviews, and three job offers. I chose to move east, and I joined the faculty at State University of New York at Oswego, a vibrant, primarily undergraduate university in Upstate New York. I was excited to start this new adventure! I wrote my first two grants before finishing my Ph.D. thesis. In my fourth year I applied for early promotion, which I was granted. At the time of writing, I will soon begin my fifth year at Oswego, and I have had such a great time working with students and my colleagues across the campus.

Suggestions for the Next Generation

I have a few suggestions for fellow and future physicists, based on my personal experiences. It is very important to have mentors: one in the same department, one in another department, and more importantly, one at another university. Since an outside mentor is at a different institution, he or she may see the picture more clearly while keeping your best interest in mind. Moreover, he or she will not evaluate you but will give you advice on how to focus, produce, and shine. He or she will remind you that the students are very important, that you should focus on writing papers and grants, and that working hard and paying attention to what is important is critical.

Another important strategy to consider, especially for those of you who start a tenure track position immediately after Ph.D.: create your own network, meet fellow colleagues, and start collaborating. One such great occasion to meet colleagues is the annual workshop for new physics and astronomy faculty, organized by APS and AAPT and sponsored by NSF.

Even if you are not evaluated every year, make sure you work on your credentials from the first month on the job. Every college and university has slightly different policies regarding the importance of teaching, research, and service. While for primarily undergraduate institutions research is very important, being a good teacher is even more important. Service does not really give one tenure or promotion, but it may be useful in order to familiarize oneself with the university and to make oneself known. Always deliver what you promise, and strive to deliver even more. Be positive and confident. Do not solely work for tenure or promotion, but work in order to make a difference in your students' lives and to give significance and recognition to your research. Keep your CV current, and make sure to keep yourself marketable. Tenure is not a destination – it is a process. If you love your job, the tenure or promotion letter should come almost unnoticed but with the clear feeling that you deserved it. Your work ethics and habits will not change after tenure or promotion (or at least they shouldn't), but those events will be a welcomed recognition of your talents and success. Take joy in the university and what it has to offer. Enjoy the students, the colleagues, and the interesting life a job in academia brings. For the stressful times, find a hobby. Currently, mine include learning a new language and how to play piano.

The support of family is very important, and I am grateful for it. Having a second child in my third year of tenure was quite taxing but simultaneously exciting and enjoyable. Some things can wait, and others cannot. With determination, passion, and daily joy you can do everything you really want! Success!!!

BARBARA A. JONES

Bona Fides

Dr. Barbara A. Jones leads the theoretical and computational physics effort at IBM's Almaden Research Center in San Jose, California. She received an A.B. degree in Physics from Harvard University in 1982, and M.S. and Ph.D. degrees in Physics from Cornell University in 1985 and 1988, respectively. She also has a Masters in Advanced Study from the University of Cambridge. After postdoctoral research at Harvard University, she joined IBM at the Almaden Research Center in 1989, where over the years she has been a manager off and on, and has won a number of internal research awards.

Barbara is a Fellow of the American Physical Society and is the 2001 recipient of a TWIN Award (Tribute to Women in Industry). She is Chair of the American Physical Society (APS) Division of Condensed Matter Physics and is a consulting professor at Stanford University. She is on the Editorial Board of Physical Review X and on the physics advisory boards of Princeton U.'s MRSEC, Georgetown U., the National High Magnetic Field Laboratory, Los Alamos National Lab, Stanford's SLAC, Canada's NSERC, and the U. of Kentucky. Barbara is immediate past chair of the National Academy of Science's Condensed Matter and Materials Research Committee.

Chair and Founder of the APS/IBM Research Internship for Undergraduate Women, member and past Chair of the APS Committee on the Status of Women in Physics (1999-2002), and past chair of the IBM Almaden Diversity Council, she is strongly interested in promoting opportunities in science and math for all students.

The Early Years

My mother was trained as a linguist but quit that and became a housewife when she became pregnant with me. My father was an industrial engineer; my father's side of the family mostly consists of accountants and engineers. My parents immigrated from England and Germany, which always made me feel like somewhat of an outsider in the Midwest. Having gotten used to this feeling from an early age made it easier, and less alienating, to later find myself a rare girl in math and science. I liked math from an early age and sought out logic and math puzzles. I liked making things from kits, logically, by following instructions. I read a lot of science fiction and other books, too, and when I was a little older, I enjoyed riding my bike far into the countryside. I loved the trains that would pass just a mile or two away from my house. Growing up in the industrial Midwest, there were lots of trains; I would hear their whistles at night – a lonesome, evocative sound.

The Awakening of My Interest in Science

When I saw chemistry and physics described in the list of high school courses, it was a revelation: I had to take those classes! They seemed so very interesting. I was lucky that my school system offered two years of chemistry, two years of physics, and math through Calculus II. I doubled up in my third year of high school and opted to take a second year of chemistry concurrently with my first year of physics. Junior year marked the first year in a newly built high school; the 2nd year chemistry class, however, was back at the old campus. There were three people, including myself, from the new high school taking this class – the other two were boys. One of the boys had a VW beetle, and he drove the three of us over to the old campus and back. The novelty of shuttling back and forth only added to the experience. My chemistry and physics teachers had opposite personalities: my chemistry teacher was very strict, while my physics instructor was rather mellow. Despite this, I learned a tremendous amount in both classes, and both classes used the math we were learning.

The popular girls in the background were not very excited about how well I was doing in my math and science classes. A number of them got together, came to me, and said, "You should not do as well as X and Y (the other two boys who I went to 2nd year chemistry with) in math and science. Let them do better; they should be the best in the class." This amazed me, but even then I recognized it was B.S. and ignored it.

My chemistry teacher told me about a Saturday morning program at local Butler University, taught by Dr. Dixon. He turned out to be a fantastic, inspiring teacher. I started going every Saturday. There I learned quantum mechanics (which I found very interesting) and also about Harvard University (recommended by a good friend in the same program). I would have never thought of Harvard otherwise; I had never heard of anyone going there after high school. My friend was accepted to Harvard, and the next year, so was I.

The Beginning of My Career

My father worked at Western Electric in Indianapolis, which had a Bell Labs location in the same building. He helped me with my resume and facilitated my securing a summer job at the Indianapolis Bell Labs after high school graduation. It was a wonderful experience, which started me on my career path, and I am very grateful to him. I worked at Bell labs in Indianapolis (with one year on the assembly line at Western Electric) until my junior year at university, and then I went to Bell Labs headquarters in Murray Hill, New Jersey for the summers after junior and senior year. At Bell Labs they always gave me meaningful, interesting projects that made use of my math and programming skills (I had taken FORTRAN and BASIC courses in the summer after my junior year in high school. I ended up using both at Bell Labs.) I also met important mentors: Drs. Chandra Varma and Michael Schluter. They worked on condensed matter/solid state physics, in particular strongly correlated magnetism, at Bell Labs, and I found I liked that subfield. They suggested advanced (graduate-level) courses to take at Harvard, and we also had very useful discussions about where I should go to graduate school and whom to work for once there.

As great as my summers at Bell Labs were, the academic years at Harvard were enjoyable and exciting, but not always easy. I had a physics professor who announced, as he was handing out exams in a class required for my major, "None of the girls will do as well as any of the boys." This was my sophomore year. I remember my reaction: "We'll see about that." On another occasion a male student told me he was switching to computer architecture because if he couldn't do as well as well as I (a girl), he had no business staying in physics. As this example shows, stereotypes can be detrimental to both genders.

I thought it would be good to spend a year in England after college graduation to meet my relatives, and I was fortunate to be awarded a Churchill Scholarship to go to Cambridge University and do Part III Maths. I had been accepted to Cornell University for graduate school, and I was able to defer admission for one year. The courses in Part III Maths were tremendously interesting and useful and included courses in perturbation methods, general relativity, and polymer theory. I found my Harvard preparation (which included some graduate courses) put me in good stead to take these classes, and at the end of the year I took the exams (which not all one-year visitors do), and passed. I use the techniques learned in the perturbation methods and many-body theory courses to this day.

The following year I began my graduate work at Cornell

153

University, located in a very rural region in central New York State. I found myself somewhat disoriented by the lack of city life. I poured my time into courses and research. My wonderful thesis advisor, Professor John Wilkins, went on sabbatical to the Institute for Theoretical Physics (ITP) in Santa Barbara for a year and brought all his graduate students with him. Before heading out to Santa Barbara in the fall, I spent one last summer at Bell Labs. Chandra Varma told me about the numerical renormalization group solution to the single-impurity Anderson model and suggested I try the same method for the two-impurity Kondo model. Chandra did not program much then, so of course I also had to teach myself the numerical renormalization group method by studying the papers. I dived in to this problem, which became my Ph.D. thesis topic, and which I took with me to UC Santa Barbara. A friend from Bell Labs told me about a house he was sharing with a few others, and I went to live with them for that year.

Because of my advanced courses at Cambridge, plus the jump-start on my thesis from Bell Labs, I finished my graduate studies at Cornell in four years. Highlights of this time were finding an unexpected second-order phase transition, with an unstable non-Fermi liquid fixed point at the transition point, as well as a new symmetry of strongly correlated models, which I termed axial charge. I then returned to Harvard to complete a postdoc, working with Professor Bert Halperin on the then-hot new topic of high temperature superconductivity. I also collaborated with two colleagues in my own generation on a $1/N$ mean-field solution to the two-impurity Kondo problem I had solved in my Ph.D. thesis, and it was interesting to see what this more approximate, but more widely used, many-body technique got right and what it missed, compared to the exact solution.

I had not been at Harvard one year when IBM Almaden in California contacted me and asked me to come out to visit and interview for a job doing theory for the extensive experimental magnetism research they were doing there, connected with the

154

disk drives IBM then manufactured down the hill. I was rather surprised by this turn in my life, since all my friends and professional contacts were on the east coast, and I had assumed that I would ultimately settle there. But I went to IBM to check it out, and I liked what I saw both inside and out. After a second visit I accepted their offer, and I have been at IBM since – over 20 years now! I have worked on a range of projects both fundamental and applied, including managing experimentalists working on the fundamental components of disk drives, quantum well theories, current-induced spin torques and other effects in magnetic multilayers. Currently I lead research to calculate the effects of magnetic atoms, in clusters or nanolattices, on metallic/insulating surfaces, as engineered and measured by scanning tunneling microscopy (STM). I also collaborate with computer science colleagues on a model of virus mutation and evolution for IBM's health sciences area. I love my work.

Reflections

I got married the year I became a manager at IBM, and I had a child a few years later. There is never really a "convenient" time to make these life changes as a working woman, but I think they are critical to overall happiness and life satisfaction. Buying a house is to a lesser degree in this same category of difficult life jumps, which I feel has many rewards. I think it is critical as a female physicist with a family to make firm rules about travel regarding the number of trips and amount of time to stay away on a given trip. I chose to cut my trips as absolutely short as possible, taking red-eye flights (which I was thankfully able to sleep on), and I was firm and consistent with all my hosts about length of stay. As my son got older I extended my travel beyond the original 24-hour limit, but now that he is a teenager, I'm seeing that it is again increasingly valuable to be at home in the evenings.

Looking back, I did not encounter gender bias either at Bell Labs or at Cornell. The senior scientists at Bell Labs, the fellow male graduate students in my group at Cornell, and my thesis advisor were all truly supportive. I think it was a very good thing to have gone to Cornell (despite the rural setting), and I am glad I took opportunities as they came early on: to take 2nd-year chemistry even though it meant a drive, to go to Butler on Saturdays even though many times I knew no one, to take the Bell labs summer job after high school graduation even though all my friends were taking the summer off, and to start living in New Jersey in the summers after junior year at university to go to Bell Labs Murray Hill (again, not knowing anyone). I think this is part of what makes success: to recognize opportunities when they come even if they are out of one's comfort zone and involve hard work. I am also glad that even when opportunities to learn physics came which would not be graded or marked "on my permanent record" (e.g. Butler University, Bell Labs, taking FORTRAN during the summer, Part III Maths at Cambridge), I dug in and did the best I could, learning as much as I could. I used every last bit of it later.

I am glad to have had a number of mentors over the years, in high school, college, graduate school, Bell Labs, and IBM. A few made romantic passes at me (and I am very glad looking back now that I said "No" every time). Some of my managers over the years have been better than others, but I think if I asked my male peers they would say the same thing. I think that attitudes toward women in physics are pervasive and mostly unconscious. They can only be combated by hard work, taking risks, and excellence, over and over. I find how I sell myself is very important. I think it is very important for senior women to support younger women coming up in the field. We have an internship program for undergraduate women, which I have been chairing for over a decade. I think if confidence in oneself and one's competence is not instilled during the undergraduate years (and earlier), it is hard to push through difficulties that arise during Ph.D. I wish we could hire more young women as interns and that other industries would take up similar programs as well. My experiences at Bell Labs were critical to my development in this field, and I hope young women interested in physics take the initiative and look for internship opportunities at universities, national labs, or in industry.

NOÉMIE BENCZER KOLLER

Bona Fides

Dr. Noémie Benczer Koller graduated from Barnard College with a Bachelor of Arts degree in Physics. She received her Master's of Science and Ph.D. from Columbia University. She worked as a postdoctoral associate at Columbia University, Physics Department, under the direction of Professor C. S. Wu. She spent the rest of her career at Rutgers University and was the first woman hired in the all-male Rutgers College, home of the Physics Department. She was the Director of the Rutgers Nuclear Physics Laboratory and served as Associate Dean for Natural Sciences in the Faculty of Arts and Science. She chaired the Rutgers Faculty of Arts and Sciences Gender Equity Committee. She is a Fellow of the American Physical Society and of the American Association for the Advancement of Science. She served as Chairperson of the Division of Nuclear Physics, the Committee on International Freedom of Scientists, and the American Physical Society's Forum for International Physics. She has received many awards and honors including the 2010 Dwight Nicholson Medal for Humanitarian Service.

How I became a physicist through war and peace, two continents, four countries, four languages, and the shifting gender barrier

Vienna, 1933 - France 1942: Early Childhood

I was born in Vienna as Hitler came to power and the racial laws began being enforced. Up to that point, my family had participated in the city's lively literary, scientific, artistic, and musical atmosphere. My father had a Ph.D. in organic chemistry, and my mother had a business degree. However, it very quickly became clear that the good times in Vienna were over and that the environment and opportunities were more propitious in France. My parents moved to Paris, and thus I had the good luck to be completely immersed in the French educational system.

The educational foci of primary school were on mathematics, language (French and Latin – to craft the mind), history, geography, and lastly, science and other languages. This emphasis changed little as we advanced into high school, but at all times, all subjects were covered. Practical experience, "social science" subjects, and sports were non-existent; on the other hand, homework was taken very seriously.

With the invasion of France in 1939, our exodus began, and we moved progressively south towards escape routes such as Marseille and the Spanish border. During this period I attended school in every town where we stopped for some time. Luckily, the French school system is extremely centralized, and the same material is taught in every French school. Therefore, changing schools didn't matter much. I actually acquired a great deal of "street wisdom" and social and political knowledge outside of school walls during our emigration.

1942 - 1951: Teenage Cultural Development

We left Europe on one of the last ships crossing the Mediterranean into Morocco and left Africa on one of the last voyages to the Americas – Cuba in our case. Havana was different from anything I had known up to that point. There was no winter, I had to learn Spanish in a new school, and at the same time, do the equivalent grade's French homework, supervised by my parents. I can't remember much about the Cuban school except its name; all I kept from that experience is a gold medal that I received at the end of the year.

In mid-1943 we were on the move again, searching for better opportunities. We landed in Mexico, a country with which I fell in love immediately, and for which I feel a passion to this day. I attended a French school, the Lycée Franco-Mexicain, and I slipped into the appropriate classes seamlessly.

In Mexico I had the good fortune of attending seven years of French high school without interruption. The teachers at the Lycée Franco-Mexicain were excellent in their disciplines and highly motivated. Many had also left France because of the war and their political leanings. The classes were small – ten to fifteen students at most. Again, mathematics (algebra, geometry, trigonometry, and elements of calculus), French, and Latin were the main foci; everything else was covered in smaller doses. During the last year, our days were filled with an almost equal mix of advanced mathematics and philosophy. We were exposed to a sprinkle of physics, chemistry, and cosmology – all taught without experimentation, demonstrations, observations, or even the notion that the subject was related to the world we live in. We had some human anatomy, but biology was nowhere to be seen. And we read and read and read . . . classics and contemporary books, biographies, history, and some science fiction.

Now how does all that lead to the making of a physicist? Well, it started with reading: Eve Curie's book about her mother Marie Curie provided a role model for all girls and filled them with the "Yes, I can" spirit many years before the phrase was coined; also fiction, history, and biographies like Jules Verne and his exotic adventures and Maurice Maeterlinck's The Life of the Bee. After the war ended, all the French existentialists' writing and ideals made it across the ocean. In addition, independence and feminism reached us early through George

Sand, Simone de Beauvoir, Marguerite Duras, Colette and others. Throughout this time, I encountered no gender discrimination whatsoever, in class or elsewhere.

Barnard College (1951-1953) and Columbia University (1953-1958): My Physics Education

After passing the Baccalaureate exams in classics and mathematics I decided to continue my studies in New York. My arrival at Barnard College, the first all-women environment to which I had been exposed, was a revelation, though somewhat disorienting. I loved New York but had trouble understanding why women were segregated across Broadway from Columbia University – I didn't really figure out the system until many years later.

I entered college as a junior, since all the humanity courses I had taken in the Lycée were recognized. I thought I would major in biology instead of the pre-med science career that my parents had in mind; however, given my background in mathematics, my adviser suggested that I major in physics, one of the most active departments at Columbia at the time. That was an extraordinarily lucky break and really good advice. I didn't really know anything about physics but loved the subject right away. Without the expert advising, I might never have chosen this field.

Immediately, and again by chance, I was hired as an undergraduate assistant in Professor C.S. Wu's laboratory. Thus began a wonderful student-professor relationship, which was both professional and personal. C.S. Wu was a young Chinese physicist who had come from Nanjing in 1936 to further her studies. She received her Ph.D. from Berkeley during the war and worked as a lecturer at Princeton and Smith College before being hired as a Research Scientist at Columbia. At the time I started, she had just finished some very important experiments elucidating the shapes of beta decay spectra and showing that Fermi's theory of the process applied, disproving previous experiments. She was an extremely careful experimentalist, had a good sense of reality, and made sure all aspects of a measurement were understood. The experimental apparatus we were using was constantly being upgraded; it was a good period to learn experimental techniques, especially for me, since my hands-on experience up to that point had been limited to fixing my roller skates and bicycle.

By 1955, some very crucial questions in physics arose, relating to two particles, which had almost identical properties but decayed via different modes. This observation led two theorists, Professors T.D. Lee (Columbia) and C.N. Yang (Institute for Advanced Studies), to question the validity of a concept that had never been questioned previously, namely whether parity, a symmetry that characterizes physical systems, was conserved in decays similar to those of the two particles mentioned and beta decay as well. As Professor Wu was an expert in the latter process, she was an instrumental part of the discussions and suggested the key experiment that would resolve the issue. Together with colleagues at the National Bureau of Standards in Washington, she carried out the experiment successfully, and the world of physics was thoroughly changed. None of the Columbia students were directly involved in the actual experiment, but we were fortunate to have witnessed years of intense activity, seminars, visitors, and planning of other experiments, observations that charged every day with new surprises and challenges. During that period I married a fellow high-energy physics student.

After receiving my Ph.D., I stayed at Columbia for another two years as a post-doctoral research assistant. In 1960, I accepted a position as an Assistant Professor of Physics at Rutgers College (within Rutgers University). Ironically, I now found myself as the only woman teaching in an all-male college! Women at the women's college, Douglass, also had to cross New Brunswick to take physics classes in the only physics department at the University. These particular gender issues were ultimately sorted out ten years later (in 1970) when

Rutgers College became coeducational, women were hired onto the faculty, and, finally in 2006, all colleges were unified into the School of Arts and Sciences admitting both male and female students.

1960 - Present: Rutgers University, and a Career in Physics

When it comes to gender issues at Rutgers, I was lucky again. The physics department in which I found myself was liberal and friendly, and I never felt any antagonism. However, it took women in other university departments many more years to overcome some of the existing biases, and environmental oversight and monitoring of the workplace climate for women is still necessary. The U.S. research foundations have been supportive, and I have been able to develop a career free of concern about survival. I have thoroughly appreciated the freedom of running my own experimental nuclear physics program.

After being awarded tenure, I had two children, both boys. My husband and I steered them towards the natural sciences: questioning, reasoning, measuring. One is now a condensed matter physicist and the other a hydrogeologist.

For twenty years or so Rutgers was home to a state-of-the-art 8.5 MV Tandem Van de Graaff accelerator. The group's unifying goal was to probe the structure of low-excited states of nuclei. The reason I liked my work so much is that it involved not only nuclear and accelerator physics but a broad range of phenomena that relied critically on the interactions of nuclei with solids, hence condensed matter and atomic physics were also an integral part of the research effort. Later, the main thrust of my work concerned the measurements of magnetic properties of excited states. These studies required the use of magnetic materials and an understanding of electronic and magnetic interplay between moving ions and the solid.

163

Eventually this research field evolved and smaller accelerators no longer provided an efficient path to exploration of nuclear matter. To adapt, we moved our nuclear physics group to other facilities such as the 88 in. cyclotron at Berkeley, the 18 MV Tandem at Yale, and the cyclotron-Tandem facility at Oak Ridge National Laboratory (ORNL). At ORNL, we were able to use beams of radioactive nuclei to study the properties of states that do not exist in nature as stable species. New equipment and techniques were developed, which were custom-designed for application at the newest radioactive beam facility, the Facility for Rare Ion Beams in construction at Michigan State University.

Cultural and Professional Activities

An academic environment provides many opportunities to contribute to interdisciplinary activities and academic management, both in the university and in society-at-large. I served on several gender issues committees and worked as Associate Dean for the Natural Sciences for several years. Through the American Physical Society, I have been involved in human rights campaigns, efforts to increase the freedom of scientists here and abroad, and promotion of international collaborations. Naturally, working to lift the ever-present gender barriers in the physics profession has led to some of my most rewarding interactions.

Reflections on a Life in Physics

All of the following have contributed to my intellectually active existence as a physicist: good schools, in which I've learned lots of mathematics and humanities; books; a nurturing family; an understanding, sharing and loving husband; good luck; and taking advantage of opportunities. A life in the laboratory surrounded by young, eager, and motivated students is a blessing. It comes at a (well-worth) cost of hard work, constant juggling of priorities, and sharp focus. But the result is a fulfilling life and the creation of a professional, personal family that is unique in our society.

LILLIAN CHRISTIE MCDERMOTT

Bona Fides

Dr. Lillian Christie McDermott is a Professor of Physics at the University of Washington and director of the UW Physics Education Group. She received her B.A. from Vassar College and her M.A. and Ph.D. in Physics from Columbia University. She is a Fellow of the American Physical Society (APS) and of the American Association for the Advancement of Science (AAAS). Among her most significant awards from physics organizations are the 2013 Melba Newell Phillips Medal, the 2001 Oersted Medal, and the 1990 Millikan Lecture Award. All three are from the American Association of Physics Teachers (AAPT). She also was honored with the 2002 Medal of the International Commission on Physics Education (ICPE). Lillian has worked to establish research in physics education (PER) as an appropriate field for investigation by physics faculty, graduate students, and post-docs. Her UW group received the 2008 Excellence in Education Award from the American Physical Society (APS) for promoting PER as a sub-discipline of physics.

Early History

I was born in Manhattan and grew up in the northern part of the island. My formal education began at P.S. 187, the public school that I attended from kindergarten through the eighth grade. In the ninth grade, I entered Hunter College High School, a public all-girls school at that time. Admission was on the basis of a competitive examination in English and mathematics. The academic program was very strong. For example, physics was compulsory for everyone in the junior year. A typing class was not even offered as an elective. (I think that there was a concern that, if we learned to touch-type, we would become secretaries.)

A partial college scholarship and a New York State Regents Scholarship made it possible for me to go to Vassar (a liberal arts college for women, then one of the "seven sisters"). My high school education had been so rigorous that I found my freshman year at Vassar a lot less stressful than did many of my classmates. I chose to major in physics, not only because I liked it, but also because I thought it was harder than other subjects and I wanted to understand it better. Developing confidence in my ability in physics, however, was a slow process. I remember telling my instructor in the introductory course that I didn't really understand the material. She kindly told me to wait until after the first exam and then discuss the problem with her. When I received the highest grade in the class, I realized that "understanding" has a different meaning for different people, (a perspective confirmed by experience in graduate school and beyond). During my four years at Vassar, the physics faculty encouraged me (perhaps because they had no male students who are typically more self-confident). When my father died suddenly at the beginning of my sophomore year, the College gave me a full scholarship for the remaining three years and, later, a small fellowship for my first year in graduate school.

Because I was needed at home, I applied only to the Ph.D. program at Columbia University, which awarded me a Higgins Fellowship for the first year (1952-1953). (Another recipient of the same fellowship that year was Mark McDermott.) Columbia differed greatly from my undergraduate experience.

Life was in the "sink or swim" mode. With few exceptions, the faculty (which included several Nobel Laureates) seemed arrogant and unconcerned about students. Two remarks (chosen from many) suffice to describe the environment. The first was made to all of the students. The second would not have been made to a man. One of the Nobelists greeted our class with the words, "Look to the right and look to the left; half of you will be gone by the end of the year." On seeing me in Mark's lab in the Columbia Radiation Laboratory, early in our second year, a future Nobelist sarcastically remarked to me that the Higgins Fellowships had at least succeeded in promoting romance. I was one of only four women in our classes to complete the Ph.D program. I was determined to survive and I had a great advantage in Mark's support. We were married in 1954.

Most of the students in the graduate program began as Teaching Assistants (TAs) and later became Research Assistants (RAs). However, I could not afford to have only a part-time salary after my first year because my mother and brother (a Columbia College student) needed financial help. I was hired as a Woman Technician (which I think was the official title) at the IBM Watson Laboratory, a short distance from the Columbia Physics Building. For about two and a half years, I clocked out of my job to attend classes. I have the unhappy memory of punching a time clock on my arrival every day. If I were even one-minute late, the printed time would turn red. There was nothing that I could do, including working late (which I always did), to avoid a reprimand. At that time, Mark and I lived in Fort Lee, New Jersey, and were part of a carpool. I remember being scolded for being a few minutes late on several days in a row. Knowing that I did not want to be treated in that way strengthened my determination to obtain a Ph.D. My final experience at IBM was even more compelling. After my brother graduated from college, I told the Director of the Laboratory of my plans to become an RA at Columbia and complete the Ph.D. program. I had thought that he would be pleased that I wanted to continue in physics, but to my surprise, he admonished me for a lack of appreciation for what the company had done for me. This episode further strengthened my desire for an academic future.

After leaving IBM, I began research in nuclear physics on the Columbia Van de Graaff. As the last of six graduate students to join the group, I had the lowest priority for machine time. By learning how to operate the Van de Graaff, Mark enabled me to collect data over weekends when no technicians were present and my fellow students were less eager to work on their experiments. I completed my research and wrote the related paper for the *Physical Review* within two years after joining the group. I defended my dissertation in Spring 1959, three weeks before our son, Bruce, was born.

In the summer of 1959, we left Columbia for the University of Illinois, where Mark had accepted a three-year postdoctoral appointment. The Department was friendly and made us feel very welcome – a great contrast with Columbia. We were therefore disappointed when, after only one year, the professor with whom Mark was collaborating accepted a position on the Columbia faculty. If Mark were to remain in Illinois, his colleague would not be able to participate in their joint project.

After our reluctant return to New York in 1960, I became an Instructor at City College of New York (CCNY). I was able to accept a full-time appointment because my mother and great-uncle volunteered to care for Bruce. I taught introductory calculus-based physics – lectures, laboratory, and recitation sections – and relearned a lot of basic physics. I also began to learn an important lesson: student enthusiasm and good evaluations are not reliable indicators of what students have actually learned.

In 1962 Mark accepted a position as an Assistant Professor at the University of Washington (UW). Although I knew that Washington State had a strict anti-nepotism policy that would preclude my employment at UW, I was optimistic that my Ph.D. from Columbia and teaching experience at CCNY would enable me to work somewhere else.[3] We moved to Seattle in September with Bruce, our three-year old son. By June 1964, he had two sisters: Melanie and Connie. In December, I applied for a faculty position at Seattle University. There were two positions: one full-time and the other half-time. I was asked which of the two I wanted. The priest who interviewed me commented that a part-time job would probably be better for a woman with three children. Since his advice seemed reasonable, that was the choice I made. In the short term, accepting a non-tenure track position was a mistake, but in the long term it turned out to be the right decision.

Early Years at Seattle University and UW (1962 – 1973)

In January 1965 I began teaching two introductory physics courses at Seattle University. I left for home at about 12:30 p.m., but I gradually became aware that some full-time faculty (all male) often left as early as 2:30. After learning that their salaries were much higher than mine, I began to question my decision to work part-time. A few years later, another disadvantage of a part-time position became evident.

During the time that I taught at Seattle U., I was sometimes asked to substitute as an instructor in introductory physics, often when the Chair realized that no faculty member had been assigned to a course. In such situations, the nepotism restriction could be suspended. By Winter 1968 I was teaching part-time at both universities. That same year, Arnold Arons left Amherst College to join the UW Physics Department. He planned to develop a physics course to prepare elementary

---

[3] Until passage of Title IX of the Civil Rights Act in 1972, anti-nepotism policies barred close relatives (in practice, mostly wives of male faculty) from faculty positions at the same state university.

school teachers to teach physical science.

When the aerospace recession (known locally as the "Boeing bust") hit Seattle, part-time faculty at Seattle U. and UW lost their positions. A billboard near the airport asked the last person leaving Seattle to "turn off the lights." Not having a job was not my main concern because the children were still quite young. I was more worried that, if I did not keep active in physics, I would not be able to return to the field. I went to see Arnold and offered to work without salary if I could help him. I guess the offer was too good to refuse because he accepted my proposal. Despite child-care complications, Mark was enthusiastic. Without his support, I could not have made such an arrangement.

I taught in Arnold's class for prospective elementary school teachers, first as a volunteer and then as a grant-supported Lecturer. I also began to develop a course for high school physics teachers and one for underprepared students aspiring to science- related careers. When Arnold was offered a position at Woods Hole, he negotiated a retention offer from UW for a new faculty position in physics education. Because the passage of Title IX in 1972 had ended the anti-nepotism policy at UW, I could apply. By this time, I had begun to build a publication record. As one of the three finalists for the new faculty position, I gave a colloquium. I spoke about the intellectual aspects of my experience in teacher preparation. I remember being terrified, but I must have done well enough because I was offered an appointment as an Assistant Professor, beginning in Autumn 1973. The salary was lower than the average for elementary school teachers.

Beginning of a Ph.D. Program in Physics Education Research at UW

By 1973 I had gotten to know three of the Department's graduate students through their role as TAs. David Trowbridge, Mark Rosenquist, and James Evans had met all the requirements for admission to the Ph.D. program but had chosen to obtain a new Doctor of Arts (D.A.) degree intended as preparation for teaching.[4] I had assumed that Arnold would be their advisor but, to my surprise, they asked to work with me.

My first teaching assignment as an Assistant Professor was the course for future high school teachers. Drawing on Arnold's book, *The Various Language,* my graduate students and I wrote worksheets that engaged students in active inquiry rather than passive learning.[5] We also began developing pretests to identify specific student difficulties. The results guided our design of instructional strategies that we assessed through post-tests. I later referred to this process as research-based curriculum development. The worksheets were the first stage in the development of *Physics by Inquiry (PbI),* one of two nationally distributed curricula developed by the UW Physics Education Group (as we had come to be known).[6,7]

Although my previous experience in preparing K-12 teachers had probably been an asset in the competition for the faculty position that I had obtained, I knew that I needed to build a research record to retain it. I was too far removed from experimental nuclear physics to do independent work in that field. I doubted that the research that I was already doing to guide the design of *PbI* would meet the Department's expecta-

---

[4] This alternative to the research-oriented Ph.D. was introduced at Carnegie-Mellon.

[5] A. Arons, *The Various Language* (Oxford University Press, NY, 1977).

[6] L.C. McDermott and the Physics Education Group at the University of Washington, *Physics by Inquiry* (John Wiley & Sons, NY, 1996).

[7] L.C. McDermott, P.S. Shaffer, and the Physics Education Group at the University of Washington, *Tutorials in Introductory Physics, First Edition* (Prentice Hall, Upper Saddle River, NJ, 2002). *Instructor's Guide* (2003).

tions. However, neither Arnold nor anyone else had specific suggestions about what I could do.

While I was trying to determine a direction for my research in 1973-1974, it occurred to me that it might be interesting to repeat with undergraduates in university physics courses the motion tasks that Jean Piaget administered during clinical interviews with students <16 years old. The interviews involved real motions that the students observed; I suggested to David Trowbridge, my first graduate student, that he ask university students the same questions. When Piaget's tasks proved easy, even for poorly prepared students, I asked David to try to develop more appropriate ones. He successfully met the challenge. Research in physics education leading to a Ph.D. in physics began in our Physics Department at that time.

David developed motion comparison tasks in which students were asked to observe and compare the motions of two balls rolling side-by-side on level and inclined tracks. He conducted individual demonstration interviews with students who had a wide range in physics preparation. Many who did well on typical quantitative problems, performed very poorly on tasks that examined their ability to distinguish the concepts of position, velocity, and acceleration from one another.

While David, Mark, and Jim were engaged in their research, the D.A. program was discontinued by UW. All three had done well, two having gotten the highest grades of all who took the Qualifying Exam when they did. It seemed to me that the Department would have no practical choice other than to grant them formal admission to the Ph.D. program, which it did. In 1979, David became the first U.S. student to earn a physics Ph.D. by conducting research in an established *PER* graduate program. Mark Rosenquist and James Evans (both outstanding students) followed in 1982 and 1983, respectively.

UW Physics Education Group and UW Physics Department

I became an Associate Professor in 1976, the first woman to be tenured in the Physics Department. In spite of the challenges that I had faced, I realized that what might have appeared to be a mistake (*i.e.,* not requesting a full-time, tenure-track position at Seattle University) had worked to my benefit. Although I had lost a year's salary during the local economic crisis, I was now conducting interesting research in a field that I would not have discovered had I chosen a full-time appointment at Seattle U. I was promoted to Professor in 1981. During the next decade, the UW Physics Education Group gradually evolved into a *de facto* Center, to which physics faculty and post-docs come to switch from traditional experimental and theoretical fields to research in physics education.

In the early 1990s, the opportunity arose to try a research-based approach to developing instructional materials for the introductory calculus-based physics course. When proposals to improve instruction in science and mathematics courses were solicited by the UW Administration, I convinced the Department that we should offer to devote one 50-minute period/week to the development of concepts and reasoning skills in small-group sections led by TAs. Our group would develop *tutorials* on topics known from research and/or teaching experience to be difficult for students. I submitted a proposal to NSF for research and research-based development of tutorials. It was funded in 1994.

The development and national distribution of *Tutorials in Introductory Physics* was a turning point in our group's history. We continued working on *PbI*, but our attention became increasingly focused on the tutorials and on the preparation of graduate students as TAs and future faculty. We have since extended our research to investigations of student learning in upper-division courses, for which we have developed tutorials

174

on more advanced topics (e.g., thermodynamics, special relativity, and quantum mechanics).

Today the UW Physics Education Group has three full professors who have supervised the Ph.D. research of 23 physics graduate students. We have also co-supervised research by three graduate students from European universities. By the end of 2012, all three had been awarded a physics Ph.D. by their universities.

Reflections on the Change in Climate for Women in Physics

My confidence in being able to succeed in physics owes much to the encouragement of my parents and to my early education in environments that were academically as demanding, but not as intellectually arrogant, as Columbia. Attending a single-sex high school and college (as both Hunter and Vassar were then) was important. Having a supportive husband, however, was the most critical factor of all.

Although today fewer than 15% of the UW physics faculty are women, attitudes have changed. For example, some of my male colleagues are quite assertive about their child-care responsibilities. Leaving a faculty meeting early to pick up a child is no longer viewed as demonstrating a lesser commitment to physics, a situation that benefits us all.

There is still a long way to go, however. The number of women who complete physics Ph.D. programs nationally remains disproportionately low. Even fewer rise up the career ladder in universities and other science-related positions. I would like to encourage others (especially young women) who might want to pursue careers in physics to develop the courage, imagination, and persistence to forge new paths towards goals that may, or may not, match their original plans.[8]

---

[8] A more complete and detailed summary of the experiences and ideas expressed in this essay are in L.C. McDermott, *A View from Physics* (to be published in 2013 by AAPT).

ANNE-MARIE
NOVO-GRADAC

Bona Fides

Dr. Anne-Marie Novo-Gradac graduated from Ohio State University in 1988 with degrees in both Mathematics and Physics. She followed her husband to Athens, Georgia, where she completed a Ph.D. in Physics at the University of Georgia in 1994. Anne-Marie started her professional career as a faculty member at the United States Naval Academy, where she taught physics and conducted research on optical materials. In 2001 Anne-Marie accepted a position at NASA Goddard Space Flight Center where she led the design team for the laser subsystem in the Mercury Laser Altimeter (MLA) instrument for the MESSENGER mission to Mercury. Designing a laser system that was required to function properly in the challenging environment of space profoundly changed Anne-Marie from an academic scientist into an aerospace engineer. During the next seven years Anne-Marie delivered two lasers for space-based instruments and led a research program at Goddard studying reliability of lasers in space. Over that time she gradually accepted more and more management responsibilities. In 2008 she accepted a position at NASA Headquarters in Washington, DC. She now serves as a Program Executive for several astrophysics missions, working with the science and engineering teams to ensure mission success.

Science: The Obvious Choice

Like most people, I credit my parents with laying the foundation for my future. My mother was an art teacher in the local public school system, and my father was a design engineer for the National Radio Astronomy Observatory. Between them, they fostered my interests in arts and sciences and encouraged me to both appreciate and to study the world around me. They also taught me to always be honest, have a strong work ethic, value people from all walks in life, and to stand my ground.

I always assumed I would attend college and get a job. Both of my parents had college degrees, and my mother worked outside the home. Science and engineering seemed an obvious choice. Most of my friends were the children of engineers and astronomers. It didn't really occur to me that their fathers were the engineers and astronomers, not their mothers. The idea that smart people become scientists rather than doctors, lawyers, or bankers was further reinforced by my social circle beyond my high school friends. My older sister introduced me to her friends attending New Mexico Institute of Mining and Technology. NMIMT is a small state-funded college that focuses on science and engineering. Not many people go there to get a liberal arts degree. Almost everyone I knew planned to be a scientist or engineer, so that is what I did too.

I was not oblivious to the fact that I had very few female classmates during my college career, and even fewer female professors. I knew that I was unusual, but I assumed that I was part of a growing population and that overt discrimination was a relic of the past. I am extremely grateful to the women who blazed the trail ahead of me over the last century. Most of my professors and all of my classmates treated me fairly. Unfortunately, graduate school was not as kind to me, and my research advisor was particularly insensitive to family issues. Four years into my studies, my husband lost his local job and had an offer for an exceptionally good job out of state. I approached my research advisor about helping me set up collaboration with one of his colleagues at a university near the new job so that I could finish my degree. He refused and told me that I needed to choose between my career and my marriage. I was unwilling to give up the four years I had already invested, and I talked my husband into passing on the good job and taking a temporary lab job so that I could finish my degree. To this day, I consider that the worst decision of my life. It robbed my husband of a great opportunity, and it kept me with an exploitive and unsupportive research advisor. In hindsight, I now realize that I would have only lost two of the four years of work, and I would likely have ended up with a far more supportive research advisor. I didn't have the courage to start over, and my family suffered for my cowardice.

NASA!

Upon completion of my Ph.D. in 1994, I accepted a faculty position at the United States Naval Academy. Although I loved teaching at the Naval Academy, the research environment was limiting. It was difficult to obtain funding, the available laboratory equipment was old and functioned poorly, and my teaching load precluded spending as much time in the lab as was needed. In 2001 I learned of a position at Goddard Space Flight Center through a friend. I was surprised that NASA hired laser physicists but was absolutely delighted to have the opportunity to work for NASA. Like most people my age, I had grown up watching NASA send astronauts to the moon. Working for NASA seemed like one of those things that only happened to other, luckier, people. Now I was one of those lucky people!

I was originally hired to help design the laser for the Mercury Laser Altimeter (MLA) instrument on the MESSENGER mission. However, two months after starting work at Goddard, the laser physicist in charge of the team left NASA for a job in industry. I was promoted to team lead for the MLA laser and given the full responsibility for delivering space flight hardware. I had been an academic physicist who spent hours tinkering in a lab alone. Now I had to lead a team of professionals and worry about issues such as mass, volume, and power consumption. To make matters more challenging, the laser was widely considered the most problematic technology in the MLA instrument. Lasers had a reputation for failing in the space environment. Luckily, I was blessed with an outstanding team who brought me up to speed and taught me what I needed to know from the NASA engineering side. My expertise with laser architecture and materials earned the respect of my team, and together we delivered a fully functioning laser for the MLA instrument two years later. MESSENGER has since completed its primary mission at the planet Mercury and is now operating in an extended mission.

I found that I enjoyed the collaborative and product-driven environment of engineering. Building hardware for space is challenging, exciting, and fast-paced. It demands solving problems that nobody else has ever solved. It's exciting to work with colleagues who have the same appetite for difficult tasks and creative solutions. My background as a laboratory physicist served me well because, like most physicists, I knew a little bit about everything. I knew enough about each discipline to understand critical design choices and their impacts on the other subsystems. The interwoven nature of the instrument as a full system is an exquisite and intricate puzzle. It takes the expertise of hundreds of highly trained professionals to make it all work. We design it, build it, test it, redesign, rebuild, retest, and eventually violently fling it off our planet into the cold vacuum of space and expect it to work flawlessly. Any one of a million things could go wrong, yet most of the time the spacecraft does the job we designed it to do. How cool is that?

The Company You Keep

My gender has seldom been an obvious issue, but it has been an issue. Most of the sexism I have dealt with has been in the form of colleagues undervaluing my expertise or my contributions. When I was at the Naval Academy, I was repeatedly turned down for a desired teaching assignment because I was "too inexperienced." The following year the class was given to a new hire, who happened to be a man. At NASA, my male team members were sometimes given credit for the work I had done. In one case I was credited with having taken the notes in the lab rather than having done the actual work. On several publications and presentations my name was listed nearly last or left off entirely. In each case, one of my male team members was given the authorship that should have been mine. Passing laws and making policies won't change this kind of discrimination. In my opinion, women doing the job and doing it well is the only thing that will change that. It needs to become "normal" for women to be capable scientists and engineers.

I want to emphasize that the vast majority of my colleagues throughout my career have treated me well. I have had the privilege of working with many exceptional people, and my own accomplishments have been greatly enhanced by having them as my co-workers. Quality colleagues share their own enthusiasm and skills and inspire you to be your best. It is key that they admire you as much as you admire them. It is not helpful to be in a situation where you feel as though you are a lowly mortal blessed with the privilege of working with gods. If you feel that way it is probably because you are being treated that way. Early in your career it is appropriate for you to be treated as a junior colleague, but if you are still being treated as a junior after your peers have moved on, it is time for you to find different colleagues.

Surrounding yourself with quality people is especially important when it comes to your boss. A good supervisor is honest, inclusive, and supportive. You should feel as though you always know what your boss thinks of you and your work. Your opinion and contributions should matter. It won't do you any good to work for a prestigious scientist if they don't include you or advance you. You should be provided with the resources you need to do the job you are being asked to do. I have had the good fortune to work for some outstanding supervisors who have made it possible for me to achieve great things. I have also had a great job turn into a terrible job overnight after the retirement of a good boss. It is critical that your boss view you as a valuable asset. You want to be the person your boss trusts with important responsibilities and to represent the organization. Of course, you should work hard and perform at a level that merits being treated that way.

Do not waste your valuable time and experience on projects or employers who do not properly support you or value your contributions. Work hard to earn the respect of your colleagues, and treat everyone on your team the way you want to be treated. Surround yourself with exceptional people. They will bring out the best in you.

LINDA J. OLAFSEN

Bona Fides

Dr. Linda J. Olafsen is an Associate Professor of Physics at Baylor University. She earned her Ph.D. in Physics at Duke University in 1997, after which she completed a National Research Council Postdoctoral Research Associateship at Naval Research Laboratory in Washington, D.C. Her research is focused on the optical and electronic properties of semiconductor heterostructures designed for laser emission in the mid-infrared. She received on Office of Naval Research Young Investigator Award in 2001, and she is currently Chair of the Materials Research Society (MRS) Congressional Visits Day subcommittee and Chair of the Book Review Board of the MRS Bulletin.

The Dual-Career "Problem"

As a woman in physics, one is already unique. As a married woman in physics, particularly one married to another physicist, there is an added layer of complexity when it comes to looking for a job. A 1998 survey[9] conducted by Laurie McNeil and Marc Sher indicated that 43% of female physicists

---

[9] "Report on the Dual-Career-Couple Survey", Laurie McNeil and Marc Sher, http://physics.wm.edu/~sher/survey.html

were married to other physicists, and that 68% of women in physics are married to other scientists. When married physicists are both seeking employment, it is often called the "two-body problem," a situation for which I would like to provide some perspective in this essay. While the "two-body problem" is a clever moniker when dual-career physicists are seeking jobs, I dislike the idea of calling it a problem, as I very much want to be with my husband. Some couples and families make things work over moderate to long distances – including bi-coastal or in some cases bi-continental situations – but that is not the path that my husband and I chose for our family.

My husband Jeff and I have been blessed to find jobs in the same location three different times. We have talked about writing a book someday to chronicle our many experiences in these job searches; perhaps this essay will jumpstart that endeavor. We have never treated the searches as one of us being the lead spouse and the other being the trailing spouse, nor have we taken turns seeking to be the primary hire. We have both done our best to find the best jobs individually and sought the best professional situation for us both, and ultimately for our family.

One Couple, Two People, Three Searches

In our first job search, I was finishing graduate school, and my husband was a postdoctoral researcher and instructor. I primarily applied for "postdocs" and industry jobs, while he applied for postdocs and faculty positions. We were engaged to be married, and we were particularly interested in locations that held promise for multiple positions, including Los Alamos, NM (where a large national laboratory is located with a diverse set of research opportunities) and near major cities such as Washington, D.C. We interviewed individually for some jobs, and in other cases enjoyed going on interviews together. When one of us had an offer, the other would look diligently in the surrounding area for a position. During this first search, I was awarded a National Research Council fellowship at Naval

Research Laboratory in Washington, DC. My husband secured a postdoctoral position in the Department of Physics at Georgetown University. Jeff was very successful, and someone asked his mentor if he would share the advertisement that he used to hire Jeff. His mentor responded that there was no ad but that Jeff had come knocking on *his* door.

Our second job search took place during the second year of very productive postdocs. I had long dreamed of working in industry, but my interest in academia had been growing. Experience as a teaching assistant, opportunities to guest lecture a variety of graduate school courses, and a good number of professional presentations had drawn me out of my shell and given me confidence in the classroom as well as in the laboratory. We both also liked the flexibility that faculty positions could offer, in terms of academic freedom with regard to our research, and hours that would allow us to care for children.

The second job search culminated in our receiving offers from the same department, part of a college in which the dean had experienced firsthand the personal and professional difficulties of dual (academic) career couples. Her sensitivity to our situation ultimately led to an accommodation that enabled the university to hire both of us. While the university had advertised one tenure-track faculty position in condensed matter, we had both applied, so the department had both our applications on hand, expediting the interview process and accommodation. We both received adequate startup packages that enabled us to build our first experimental labs.

We worked furiously as Assistant Professors. During this time I brought in over $1 million in funding, and my husband published in journals that included *Physical Review Letters* and *Nature* (to name just a few of our professional achievements). Despite this success, the department chose to endorse tenure for only one of us. Naturally, we did not agree with the decision, and we decided to look elsewhere for faculty

185

positions.

In our third job search, which occurred over two years, we were fortunate to end up with multiple offers from which to choose. Searching for a faculty position for a second time is quite different than doing so the first time. We had established ourselves more professionally by that time. I was "more comfortable in my own skin," and while we earnestly desired new tenured/tenure-track positions, we were not desperate and could be very honest about our experience, aspirations, and opinions. In this case, we were hired into a department that was advertising for two positions, and the culture of the department and university was generally more collegial than that to which we had sadly grown accustomed at our previous institution.

Lessons Learned

One should note that we never limited our applications to the institutions or cities that advertised more than one position – often, a second position would be advertised later, or other opportunities would arise after the original application deadline. Neither did I limit myself strictly to advertised positions in physics – while I am trained as a physicist, my work in semiconductor lasers has overlap with electrical engineering, materials science, and chemistry. So we cast a very wide net. Of course, if a place did advertise two positions, especially ones to which we were particularly well suited, we jumped at such opportunities.

It is illegal for employers to ask about your marital status. Many experts, and perhaps other authors of essays in this collection, will advise you *not* to inform your potential employers about your spouse until after you receive an offer, or perhaps not until you accept the job. However, we made a very deliberate choice to be forthright about our marriage and professional interests and that we were seeking positions in locations nearby. While sometimes this forthright approach led to awkward discussions, it also led to many honest conversations about the direction of the department or program, or the availability of faculty or industry jobs. As a couple, we did not want to waste our time or the institution's time or resources if there was no possibility of us both ending up with jobs in the same vicinity.

We often both submitted applications for the same position, and since I took my husband's surname, it was obvious to *most* departments that we were both seeking jobs. In one notable exception, a potential employer met me between sessions at an American Physical Society meeting to discuss inviting me to campus for a formal faculty interview. This individual and his department had conducted phone interviews with both Jeff and me; however, he was shocked when I asked about potential openings and opportunities for my spouse, suggesting that I was not interested in even interviewing unless they first offered my husband a job, when in fact I was just querying about the possibilities in the coming years to see if this was a viable option for us – again, before wasting valuable time and resources of my own and of the institution. Fortunately, his overreaction was the exception rather than the norm, but we found that most potential employers valued that we were very up front about our personal and professional aspirations to secure jobs in the same locale.

Finding two positions in physics can be quite daunting, but there are some advantages, including additional flexibility of the department or dean to create a second position in the same department. While a physics department may believe you are the best thing to come along since sliced bread, they may have very little influence if your husband is seeking a job in English. However, if they are really interested and impressed with you, it is easier for them to work with both you and your husband if he is also a physicist.

At one point in graduate school, my then-friend and now-husband swore to me that he would never date or marry another physicist, as he felt like he would never be able to leave work. One of the aspects of our professional lives that works very well is that we have different research specialties. Our experimental expertise is complementary, so we can provide valuable assistance in the laboratory. We are each fascinated by what the other does without feeling the need to pursue our spouse's specialty. We have established our professional careers separately, but when there is a project of mutual interest, we are not afraid to pursue it. Do we have trouble leaving work at work? Sometimes, but more often that is tied to politics or administration rather than to research. Before having children, Jeff did not need to worry about me being upset with him for staying in the office until the wee hours of the morning working on a grant proposal, as I was doing the same a few doors down the hall. And now it is easier for us to let work go in the evening hours as we are blessed to spend them with two beautiful young children – one of whom already appreciates the potential danger of Mommy's laser or the high magnetic field region in Daddy's lab.

There is no single or "correct" path in finding jobs as dual-career couples, whether or not both spouses are physicists. This essay highlights the results of my experiences in three separate job searches. In each of these cases, the path to securing two jobs was different. We have worked at separate institutions and also in the same departments, and there has been great value to finding those jobs and to serving in all those positions. Whether you speak extensively with other women physicists and dual-career couples as you prepare for your job search or independently forge your own path, know that there are many examples of those who have come before you and succeeded in securing positions together. And never underestimate the importance or value of finding a husband who will support your work as much as his own.

ANGELA V. OLINTO

Bona Fides

Dr. Angela V. Olinto received her B.S. in Physics from the Pontifícia Universidade Católica of Rio de Janeiro, Brazil, graduating at the top of her class. She received her Ph.D. in Physics from the Massachusetts Institute of Technology (MIT) in 1987. She was a postdoctoral fellow at the Fermi National Accelerator Laboratory (Fermilab) in Illinois and subsequently became the first female faculty member in the history of the Department of Astronomy and Astrophysics at the University of Chicago. Now a Professor in that department, in the Enrico Fermi Institute, and in the Kavli Institute for Cosmological Physics, Angela is the U.S. PI of the JEM-EUSO space mission and a member of the international collaboration of the Pierre Auger Observatory, both designed to discover the origin of the highest energy cosmic rays.

Angela served as Chair of the Department of Astronomy and Astrophysics from 2003 to 2006 and began a second term as Chair in October 2012. She is a fellow of the American Physical Society and of the American Association for the Advancement of Science, was a trustee of the Aspen Center for Physics, and has served on many advisory committees for the National Academy of Sciences, Department of Energy, National Science Foundation, and the National Aeronautics and Space Administration. In 2006, she received the Chaire d'Excellence Award of the French Agence Nationale de Recherche, and in 2011 she received the Llewellyn John and Harriet Manchester Quantrell Award for Excellence in Undergraduate Teaching at the University of Chicago. She is an advocate for increasing the participation of women in science and has advised a significant number of women graduate students and postdocs during her career.

The Early Years

My childhood was marked by a mixture of city life in Rio de Janeiro and summers spent horseback riding in the mountains near Rio with my grandparents and their numerous grandchildren. After my parents divorced, my mother moved to Brasilia with my brother, my sister, and me. This move marked a major change in my worldview. Encouraged by an urge to return to Rio, away from the foreign environment of the military dictatorship centered in Brasilia, I enrolled in an experimental program, which allowed me to finish high school in two years instead of the traditional three. Very few students were able to finish the double pace of classes, but I succeeded, passed the national exam to enter the Pontifícia Universidade Católica (PUC), and returned to Rio de Janeiro – all at age 16.

Thanks to the help of my father, who spent a month teaching me the necessary math that I had not learned in Brasilia, I performed well on the entrance exam and was placed in a special class for the top students entering PUC in 1978. PUC was considered home to the best physics program in Rio – and maybe in the country – at the time. However, I was not sure if I would major in physics. I also liked architecture, economics, and history (although my ability to remember names and dates was always much worse than solving mathematical problems). Entering the physics program was harder than the other options, so it made sense to start there and transfer later if I did not like it.

I loved it. It was a great program with small classes, excellent teachers (most having Ph.D.s from U.S. universities), and wonderful students (who quickly became my friends). The atmosphere in the country was that of an imminent revolution, which kept me as busy leading the student organizations against the military dictatorship as learning electromagnetism and quantum mechanics. Both turned out to be useful skills for my future role in science leadership.

In the last semester of the 4-year undergraduate program, I became very ill after finishing my TOEFL and GRE exams (required to apply for Ph.D. programs in the U.S.). I eventually learned that I had a rare autoimmune disease, named polymyositis, which kept me bedbound until mega doses of medication allowed me to survive and finish my studies. I was thrilled to be accepted into the Ph.D. program at MIT (my first choice), but it took about a year of support from my family and doctors plus a visit to the Mayo Clinic with my mother to become strong enough to join MIT in the fall of 1982.

Physics in the U.S.

I arrived on campus very weak and would not have been able to begin my Ph.D. if it were not for the support of my husband and MIT. I met Davi Geiger, my first husband, in my first year at PUC (he was 1 year ahead of me in the physics program). I persuaded him to join me in the graduate physics program at MIT, and he supported me through the difficult recovery phase. He is now faculty at New York University and the Courant Institute in Computer Science.

MIT helped us find a studio apartment and a doctor who knew this rare disease, both on campus. MIT also gave me the first glimpse of how male-dominated the field of physics was in the U.S.: in one of our largest graduate classes there were 60 graduate students but only 3 women. This was a surprise to me. At PUC we had women teachers, and about 30% of the students were women, enough to not consider gender an issue. The combination of a highly competitive environment and the total lack of role models made me very aware that my gender may be perceived as inappropriate for the field; this defined my first encounter with the "impostor syndrome." I felt like I was a fake despite external evidence of the contrary, such as the best grades in the class or positive recognition of my work. It did not help that some fellow graduate students looked at we women as if we were aliens, which made for several awkward encounters in those cold-looking hallways, offices, and through the nascent internet.

Aside from good friends in Cambridge, MA, what helped keep me sane at MIT was the idea that my husband and I would return to Rio after completing our Ph.D.s – we had no interest in competing for positions at U.S. institutions. My father had done his Ph.D. in physics at MIT and had returned to a leadership position in Brazil when I was 2 years old. My mother did not understand why her (ex-)husband or I had to study so much and kept suggesting I open a restaurant instead. So I mostly had self-imposed pressure to do well, stemming from my tendency to enjoy a good challenge.

The plan to return to Rio was postponed by my divorce and by the exciting research project that led to my thesis. I specialized in particle physics theory at MIT, taking the three stages of the general exam (a memorable experience) and choosing particle physics theory for the oral exam. My thesis project, the study of a quark matter state named strange matter, started out as particle physics but quickly introduced me to astrophysics. I was lucky to work with two excellent advisors, Eddie Farhi (my official advisor who tried hard to thicken my very thin skin) and Charles Alcock (who taught both Eddie and I all the astrophysics we managed to grasp in that short time).

My Ph.D. thesis, entitled Strange Stars, and the papers published on that work were very well received (still a highly cited work) and landed me a postdoc position in a very vibrant new group of particle astrophysics theory, led by Rocky Kolb and Michael Turner at Fermilab. The year I graduated, I also met my second husband, Joshua Frieman, who joined me at

Fermilab as a junior faculty (he is now a leading cosmologist and faculty at both Fermilab and the University of Chicago).

Establishing a Career

At Fermilab, I began trying to establish myself as an independent researcher. The field of particle astrophysics was (and still is) fertile ground for taking on new problems. From strange stars to inflation, there was a wide range of new phenomena to investigate. I tended to avoid very crowded subjects and did not want to repeat much of what I had done in the past. I kept some interest in the very novel idea of strange stars but realized that there was no clear evidence for their existence (or inexistence), and it may take decades before we could glimpse the internal structure of neutron stars to the needed level. I joined forces with Josh and Katherine Freese, proposing a new model of inflation and started a long-term effort to study cosmological magnetic fields.

At the invitation of David Schramm, I joined the University of Chicago, first as a Senior Scientist and then a faculty member. Schramm was a great visionary, leader, and friend who pushed the frontiers of particle astrophysics and the leadership of the University of Chicago. At Chicago, I began advising graduate students and postdocs who strengthened my research program as I chose good projects for them to succeed. Focusing on them helped me through many difficult situations that arose as the first female faculty member of a large all-male department. Most of my colleagues were welcoming, but there were also a number of inappropriate comments, misconceptions, and prejudices that did not help my ability to do science.

By the time I became faculty, I had gone through a second divorce. My third (and last) husband, the classical guitarist Sergio Assad, accompanied me through the tenure track process. He travelled a lot for his concerts, so I took on the responsibility of taking care of his three teenage children from two different countries (Brazil and France). It was a colorful, crazy household to be in while tackling the challenges of tenure track without any help other than a cleaning lady once a week. Somehow we all survived this phase and have succeeded in finding our ways. I have a fuzzy memory of how I juggled it all. I remember finding much comfort on the writings of role models – pioneer women that took on a much larger share than the average person. The chaos I had to constantly avoid at home also helped tame my tendency for perfectionism and made me more productive as a scientist. I had no time to let the impostor syndrome or other distractions kick in and stall my research efforts.

Advantages of Aging

Looking back at my past achievements I feel a sense of accomplishment. My work has been well-recognized, and my past students and postdocs have moved on to live good and productive lives. I am currently starting a new effort to advance my chosen subfield, which I hope will bring answers within the next decade to century-old questions, a possibility that is a source of great energy to me.

My path to this stage was not straight and has often involved ignoring – as much as possible – uncomfortable situations. I describe a few of these here to help those that may be faced with such unfortunate moments, which are even more likely to happen to members of minority groups in physics (like women). For example: a professor once commented on my figure just before introducing me as a colloquium speaker; another invited me to a conference and decided, without my knowledge, that I should stay at his house instead of the hotel (while his wife and children were not there); and a seminar that was introduced by "I do not believe anything that she is about to tell you." I also remember cooking a nice dinner for some senior figures in theoretical physics, and then having to argue that studies of ant populations should not be used as a proof that women cannot do theoretical physics, only experiments. Commonly, one hears the jealous comment, "she only got the job because she is a woman," but our opinions are not as commonly heard or even asked for. There are complaints about our clothes being "distracting" and colleagues who introduce the rare woman colloquium speaker by praising her husband. These are examples of some of the difficulties of this path, and women are still likely to run into many biases, but fortunately cultural change is happening, however slowly.

The awkward moments I mentioned above were spaced few and far between, which gave me time to calm down and "keep my eye on the ball." However, these and other situations encouraged me to behave more as an outsider in the field. One consequence of this choice is that I may be the last one to learn of some relevant development in the field – information which travels faster over a beer, a hike, or a squash game with colleagues than through publications. The outsider role does have some advantages however, as it brought many diverse people into my life – musicians, artists, philosophers, business folks, simple folk, bungee jumpers, and skydivers – more than just scientists. That said, it's worth pursuing some balance, as being an insider may help advance a scientific program and help you learn the "law of the land" faster.

Some Specific Suggestions:

- Don't apologize. As a child, I used to infuriate my friends by saying sorry every time I missed a serve. The same principle applies with talks or discussions. No apologies, just play the game, and keep your eye on the ball.

- Beware of the impostor syndrome. Don't compare yourself to science idols such as Einstein or Feynman. Look around you and give a more balanced assessment of the work you are doing. Compare yourself to those in the same stage of the career. If you are one of the top students, you are doing great. If not, then try to become one.

- Be patient. Science is hard work, but the thrill of a new idea or a new project that may actually pan out makes all the work worth it.

- Enjoy the new ideas, even if they are not yours. If it really interests you, then run with it – you may make it more real then the first person who glimpsed at a piece of truth.

- Practice explaining your work to non-experts in concise and creative ways. You may need it as you become one of the few experts in a topic that very few of your colleagues can follow.

- Travel the world and meet people from different cultures who all know quantum mechanics and Einstein's equations. You may become convinced that science is one of the greatest monuments that our civilization has built.

MARJORIE A. OLMSTEAD

Bona Fides

Dr. Marjorie A. Olmstead is Professor of Physics and Adjunct Professor of Chemistry at the University of Washington (UW), Seattle, where she currently directs the interdisciplinary Nanotechnology Dual Ph.D. Program and is Associate Chair of Physics for undergraduate affairs. Before joining the UW faculty in 1991, she was an assistant professor at the University of California, Berkeley (UCB), and a faculty scientist at the Lawrence Berkeley National Laboratory. At both UW and UCB she was the second female physics faculty member in the department. Between receiving her Ph.D. from UCB Physics (1985) and joining the faculty there, she spent 18 months as a member of the research staff at the Xerox Palo Alto Research Center. She received her Bachelor's degree in Physics with highest honors from Swarthmore College in 1979 and spent the summers of 1978 and 1979 at Bell Laboratories in Murray Hill. Marjorie's research centers on the formation of interfaces between dissimilar materials and the intrinsic properties of the resultant nanostructures, with particular current interest in materials that exhibit intrinsic vacancies in their crystal structure. She received an NSF Presidential Young Investigator Award (1987), the 1994 Peter Mark Memorial Award of the American Vacuum Society (now AVS) and the 1996 Maria Goeppert Mayer Award of the American Physical Society (APS), and is a fellow of both the APS and AVS. In 2000 she received an Alexander von Humboldt Research Award, and she has twice received the UW Society of Physics Students Undergraduate Teaching Award.

She chairs UW's Faculty Council on Women in Academia, which is currently focusing on documenting gender-based experiences of both ladder and non-ladder faculty at UW and addressing issues raised by these efforts.

Early Life in New Jersey

I can't remember a time I wasn't going to be some sort of scientist or engineer. My mother claims she knew I would be an experimentalist when as a toddler I dismantled a piece of furniture she didn't know had parts. I followed the "standard path" for physics with no diversions – good suburban high school, liberal-arts college, two summer jobs at Bell Labs, graduate school, industrial-lab postdoc, and then faculty position by the age of 28.

There was never any question of my aiming for college, or of my being able to pursue whatever career I wanted, whether it be "smartest housewife on the block" or engineer. All the adults I knew had gone to college, and all my friends were aiming at (and most of them went to) elite colleges. My grandmothers were both teachers until their children were born, my maternal grandfather was a mechanical engineer (Stevens Institute of Technology, 1912) and my paternal grandfather had a Ph.D. in physics (Princeton B.S. 1919, Ph.D. 1923), retiring from a forty-year career at Bell Labs as I started kindergarten. Their technical degrees meant they each kept a good job throughout the depression, and as a physics student during WWI, my grandfather was put to work calculating artillery trajectories in Maryland rather than getting fired upon in Europe. Everyone close to me supported my career goals.

My mother's degree was in chemistry (Bryn Mawr 1950), and she initially worked as a technician at Johns Hopkins Medical School. When she moved to New Jersey in 1952 to marry my father (Princeton psychology, 1949), she interviewed at what is now Ciba-Geigy, where she was told, "We don't hire young women who just got married, since as soon as we train them they get pregnant and quit." She told me this story as I was heading out for my first job interview saying, "At least they can't tell you that anymore. They can think it, but they can't say it." Ciba-Geigy hired a male chemist instead who stayed for less than a year, using the position as a stepping-stone to another company in another state. My mother worked in a hospital pathology lab for three years and lived in the area for another 35 years; she might well have remained a practicing scientist after my sister was born had she been offered the more interesting job at Ciba-Geigy. Instead, she became a "professional volunteer," active in scouting, AAUW, PTA, the town environmental commission, and church. My sister (Franklin and Marshall geology, 1977) worked for five years as an exploration geologist for a coal company, where gender discrimination was rampant, and never went back to work after a merger led to all the women (and none of the men) in her department being laid off. My mother's and sister's experiences made me both determined not to let external forces control my choice of career and appreciative of the value of "professional volunteering."

Transition to Researcher: College

I thought I was going to be a nuclear engineer, and I chose Swarthmore College as a liberal arts college that offered engineering; however, by the end of the first semester I was a physics major. The first clue was when I got a poor grade in Engineering 1 on a design assignment, despite a good design, because it wasn't presented in the proper format. Engineering 2 (statics) cemented my belief that I was not an engineer: I liked deriving formulas, not "plug and chug." At sophomore registration (which was in person, with all the faculty and students in the dining hall), the electrical engineering professor told me he didn't like having bored people in his class and that I should instead broaden my horizons. I ended up taking an upper division music class on Mozart; I worked harder for and was prouder of my "B" in Mozart than any of my A's in math or physics. Two years later, I interviewed at Bell Labs with a scientist who would later win the Nobel Prize; when I walked into his office the first thing he said was, "Tell me about that B in Mozart." We ended up with a great discussion about the benefits of a liberal arts education. Other Swarthmore skills that served me well in later job interviews were willingness both to admit ignorance and to ask challenging questions. An IBM scientist once told me my challenge of his data interpretation proved I would be a useful colleague – he was tired of interviewees that only told him, "Yes, Dr. A. Whatever you say, Dr. A."

In my junior year I applied for summer programs at Argonne and Brookhaven, as well as at Bell Labs, which paid more than the national lab programs and was close enough to live at home. After taking nuclear physics while my applications were being reviewed, I was no longer enamored with it. I thus pursued semiconductor physics in the "Summer Research Program for Minorities and Women" at Bell, which was a fantastic experience that shaped my career. I also returned the following summer. I learned to read the literature, to deal with recalcitrant equipment and, especially, that while anyone can take data, the scientist is the one who figures out what it means and plans the next experiment. The summers at Bell also taught me about the culture of science beyond the sheltered, gender-neutral Swarthmore environment. It was hard not to notice that essentially all the women and minorities were summer students or clerical staff; I also noted distinct segregation in the Bell lunchroom between the Ph.D.s (who read the New York Times) and the technicians (who read the Daily News). The main things I got from that summer, however, were confidence in my experimental abilities, an appreciation for condensed matter physics, and the knowledge that the kind of job I wanted required a Ph.D. Also, while I have met very many people (including physicists) whose stereotype of "physicist" didn't include "female," I have met very few physicists who didn't treat me as a competent scientist once I had talked physics with them for a while. I view these special programs for under-represented groups as giving people the chance to develop the "talking physics" skills that allow them to be accepted in a larger context.

Moving West: Graduate School

Graduate school at Berkeley was a marvelous experience. The physics was exciting, the city was vibrant, and I met my future husband the weekend before classes started. I was forced to do some real soul searching when picking a research group, since my first choice was to work with a non-faculty-member at the Lawrence Berkeley Lab who had been attracting "too many" good students. I talked with both physics faculty and my Bell Labs mentor, made a big list of pros and cons for each of several research groups, and eventually convinced the department chair to let me join the group at LBL. While I resented it at the time, it was a good experience in retrospect, since I was prepared for some problems that did crop up, developed fruitful mentoring ties with physics faculty members and learned the importance of being on good terms with the department chair.

When my incoming class arrived at Berkeley with 5 women (out of 65), we increased the number of women in the program by nearly 50% (to 16). There were no women on the faculty. The female students would meet for a potluck once a quarter. We mostly talked about issues that the men students had to deal with as well: quals, prelims, and finding advisors. Once, though, someone asked the group how many of them had ever cried in front of a professor and well over half the hands in the room crept up. We were all amazed (and empowered) not to be the only one. As a professor, I always keep a box of Kleenex on my desk (and, yes, the male students cry, too).

The male faculty wanted to be supportive, but most weren't quite sure how. The Electricity and Magnetism professor tried

201

by assigning an essay for homework on why there were so few women – needless to say, that didn't help. Another asked me at tea, "What's different about being a woman in physics?" to which I replied, "You get asked questions like this." Fortunately, some faculty knew what to do. The department chair, Dave Jackson, was committed to hiring a woman faculty member, and eventually hired Mary Kay Gaillard and her husband, both particle theorists. That spring, the women students invited Dave Jackson to our potluck and gave him what we deemed the highest award we could think of – a certificate naming him an "honorary woman," which we all signed. I am very grateful to Mary Kay for "walking on water" successfully, so that I could get my ankles wet when I joined the faculty a few years later.

Entering the "Real World"

One huge advantage of grad school at Berkeley was that recruiters for big research labs paid regular visits. When I was ready to go onto the job market, I knew them well enough to call to set up interviews. I was pretty sure I wanted a faculty job eventually, but at the time the "standard path" in semiconductor physics included a postdoc at one of the big-name industrial labs: Bell, IBM or Xerox. I chose Xerox Palo Alto both because I could work at a local synchrotron, adding important skills to my experimental "bag of tricks," and because my then-future husband had a lectureship and post-doc lined up at Berkeley while Bell and IBM were on the east coast.

After I gave my thesis-defense/job talk as part of the condensed matter seminar series at Berkeley, I got a mysterious message that the department chair wanted to talk with me. He told me two of my faculty mentors had decided my talk was better than those they'd seen from faculty candidates and that I should apply for the open assistant professor position. I was flabbergasted, and I protested that I wasn't yet ready. He said I could defer for a year to pursue my postdoc at Xerox if I got the offer, but there might not be an opportunity for a long time if they hired someone else in my field. It took some major convincing by my friends for me to believe in myself enough to apply, but I did. I never had an interview, and was never told officially that I was on the short list. Eight months later (May 1985), while I was in a meeting where Xerox managers were detailing their new, more application-driven vision for their research lab, the Berkeley physics faculty voted to make me an offer, with a starting date of July 1986. UCB hired five people that year – three of us were recent UCB Ph.Ds.

When I arrived back on the UCB faculty I was half the median age (28 *vs.* 56). The older faculty members weren't quite sure how to treat me, but I think "daughter" comes the closest. I never really felt like the "local expert" on anything until I moved to Seattle a few years later. There were numerous department social events, including one that required formal dress (good thing I had saved a bridesmaid's dress). The wife of one eminent faculty member sought me out at a cocktail party to tell me, "I'm so glad those old fuddy-duddies finally hired a young woman." I always felt awkward at these events, as it was never clear whether I should hang out with the women or the faculty; my future husband very rarely attended them, as he didn't have much in common with either group. I once wandered from a cluster of wives into a group of faculty and heard the phrase "affirmative action" from one. I asked, "Are you talking about me?" and he replied, "Oh, no, Marjorie. You're good."

Adding another woman to the faculty raised structural issues: neither my office nor my lab was on a floor with a ladies room. There had been female graduate students working in the basement and female secretaries on the third floor for a long time, but the department never acted to remedy the problem until I joined the faculty. The basement floors were easy, since there was only one toilet anyway: they just put a lock on the door. I'm told the grateful female grad students dubbed it the "Olmstead Memorial Bathroom" after I left. The 3rd floor bathroom, though, had three stalls and a urinal, and the male faculty initially gave efficiency-based arguments about losing access when it would be only occupied by one person. Then one colleague spoke against change with, "Besides, I have my best conversations with the chair when we're in the john." I replied sternly, "That's the point," whereupon they realized they needed to do something. The compromise was that women would lock the door and men would set the lock before closing the door so other men could come in but women would know it was occupied. Most of the time they remembered to flip the lock back ….

Establishing a Career

Within a few months of my arrival on the UCB faculty, high temperature superconductors were discovered, and a very large fraction of condensed matter physicists hopped on that bandwagon. For me, it was wonderful since my competition all switched out of surface and interface science for a while, giving me the time to get my new laboratory up and established. A few years later, nanotubes and buckyballs blossomed about the time I was moving from UCB to the University of Washington, helping distract the field once again. I learned from this the advantages of not working "boson problems" that the whole community condenses into, but rather on interesting physics in less-common materials where students can take a couple of years to get hold of the problem without being scooped.

During my third year on the faculty, my future husband, who was then a post-doc 3000 miles away at Yale, started applying for faculty jobs in cell biology. Berkeley had not yet figured out how to deal with dual-career couples (although I believe they have by now); most faculty could not fathom why anyone would give up a professorship at Berkeley for a less prestigious institution. They lost five young faculty from physics and biology around the time I left for dual-career reasons – all of whom and their partners went on to successful careers at other institutions. University of Washington, on the other hand, had a proactive provost and a supportive physics chair who provided the backing for both zoology and physics to make multiple offers that year. I effectively invited myself to my interview: once my future husband learned he was in the final three candidates for zoology, I called the physics search chair to say I would be in town visiting friends and would like to stop by the department. They were almost done with the search process, but he hastily set up a talk and interview once he learned I would be willing to come without tenure and was paying my own way to Seattle anyway. We moved to Seattle in Jan 1991, I got tenure in Sept 1993, and we were married in July 1994.

Moving Forward

Since arriving in Seattle, my life and career have gone well with remarkably few barriers. There were some political struggles and sleepless nights resulting from the students and shared resources I left behind in Berkeley, but those dissipated once the students graduated and I got new funding in Seattle. I had great support from my chair and colleagues while dealing with pregnancy, maternity leave, and a major illness in my family. I've been offered more leadership positions than I chose to take on, and I have learned to "just say no" to extra responsibilities that I don't have time for. Dispensing 24 hours a day into competing demands of family, teaching, research and service requires choices. When a new opportunity comes along, I only take it if something else can be shed. This can be giving up an existing research direction to follow a new idea, negotiating the end of departmental obligations to make room for ones at the university level, or saying no to an invited conference presentation to make room for chaperoning a school trip. Overall, I'm pleased with my choices.

MICHELLE L.
POVINELLI

Bona Fides

Dr. Michelle L. Povinelli joined the USC faculty in 2008 as an Assistant Professor and holds the WiSE Jr. Gabilan Chair. She is the recipient of an Army Young Investigator Award, a NSF CAREER Award, and the Presidential Early Career Award for Scientists and Engineers. In 2010, she was selected for the TR35 award by Technology Review Magazine, which honors top innovators under age 35. Michelle earned a B.A. with Honors in Physics from the University of Chicago, an M.Phil. in Physics from the University of Cambridge, and a Ph.D. in Physics from the Massachusetts Institute of Technology. She did postdoctoral work at Stanford University in the Department of Electrical Engineering. She was awarded several graduate fellowships, including the Lucent Technologies GRPW Fellowship, the NSF Graduate Fellowship, the MIT Karl Taylor Compton Fellowship, and the Churchill Fellowship. In 2006, she was one of five national recipients of a $20,000 L'Oréal For Women in Science Postdoctoral Fellowship award. She has co-authored more than fifty journal articles and co-invented three US Patents.

Growing Up

I didn't do any of the things physicists are "supposed" to do as kids: fix cars with their dads, take apart radios, or play with chemistry sets. I was a different sort of kid. I spent all day reading fiction, devouring a novel every day or two, and taking occasional breaks to play sports (poorly).

It is not uncommon for one of my male physicist friends to report that reading *Surely You're Joking, Mr. Feynman!* inspired them to continue on their path in physics. I read this book when I was an aspiring physicist as well. However, my main reaction was that I had no interest in cracking safes, nor in doing figure drawing of nude women in my home.

Despite this seemingly deficient background, I have nevertheless grown up into a reasonably successful scientist. In fact, my love of narrative has served me well. My job is to write stories: the story of how I got interested in a particular topic, what I learned by investigating it, and what that suggests for future investigation. I am not a technique junkie. My journal articles are rarely technical *tours de force*. I believe in asking interesting questions, in debating and analyzing and ripping apart concepts, only to put them together again. The particular techniques will follow. When I am not working, you might find me talking about developmental psychology, documentary filmmaking, or best practices of weightlifting.

My point is that there are different paths into science. Maybe you are a person who builds things; maybe you are a person who tinkers with your hands; or maybe you are a person who is just interested in why and how things work, and how to use this knowledge to solve practical problems. There is room for all of these people in science.

Valuing Your Own Goals; Proper Care and Feeding of Advisors

If you are going to get into the science business (particularly academic science), you have to realize that it is all about your own intellectual development. It is certainly not about the money. There is a certain tendency among students, once the excitement of the first year wears off, to start acting like indentured servants. This will not do anyone any good. No matter what anyone else around you is doing, you must keep asking yourself: *What do I want to do? What do I want to learn? And where do I want to go next?*

A key aspect of graduate school is your relationship with your advisor. In an ideal world, she will be an inspiring, god-like figure with an amazing perspective on science, excellent technical knowledge, and an uncanny ability to assign projects that result in papers published in *Physical Review Letters*. In an less-than-ideal world, you may find yourself stowed in the corner of the lab, taking orders from a megalomaniac senior grad student, and wondering what exactly your advisor does, since she hasn't stepped foot in the lab in the last 3 months.

Here's how it looks from the other side. Your advisor wants to do science. She has a set of topics, questions, and techniques that she is interested in exploring and developing. She also has practical constraints: she needs to raise money to keep you fed, plus buy all those expensive semiconductor wafers you need for your microfabrication project. She is constantly reporting to her bosses – program managers at the funding agencies, academic mentors, and the people who decide on tenure or promotion or career advancement. She wants *you* to develop… and she needs you to get *her* stuff done. You should figure out how to make your interests align. Some students make the mistake of listening too much to their advisor, thinking if they just do everything she says, they will end up with a Ph.D. It doesn't work this way. Your advisor, no matter how smart, does not have the time to figure out your Ph.D. thesis for you. On the other extreme, some students feel so constrained by taking direction that they hide from their advisor for months at a time, in the hopes of doing whatever they want. This is also a bad strategy. Your advisor is an important resource. All that time she is not in the lab? She is out there in the world, networking with other scientists, trying to figure out how to put her lab's work in perspective of the larger trends in the field. She is thinking about what problems are most important and interesting to solve and how to market your results to the world when you do solve them. Keep the lines of communication open and communicate frequently, and she will help you turn your results into papers, which are critical currency if you want to continue in science.

Developing Self Discipline

When I was an undergraduate student, I got fired from my experiment. Or at least, that's how I like to tell the story. I was taking data on granular compaction, and we were studying processes that happened to have exponential time dependence. Practically, this means I had to record the height of a vibrating sand column at ever-increasing intervals. The first few minutes were not bad (relatively frequent data taking), and the last hour or so wasn't bad either (infrequent intervals – time to check email or work on a problem set!). In between was sheer torture. What do you do for 1.6, 3, or 7 minutes? (This was in the days before Facebook.) Suffice it to say, one Thursday night, my experiment broke, and I was not entirely displeased. I left it idle for the weekend, untouched, and when I came in Monday, my advisor informed me that the postdoc in the lab would be taking over.

I was crushed. Further inquiries (Was it my fault? What should I have done?) resulted in a long speech about how if I didn't *like* my project, or the *people* I was working with, or my lab, perhaps I needed to think about where my *real* interests lie! I promptly became a theorist. I figured the machines wouldn't break. Ironically, I am now a theorist with a lab that does experiments. With machines that break. Don't get me started about the silicon etcher.

This unpleasant speech was an important step in my career development. Ever since then, I've learned that I need to push my own projects – relentlessly. You are the only one who is going to get it done. Either that, or you may find yourself replaced by a postdoc.

I hope you do not have the problems with self-discipline that I had. If you are lucky, you and your work are one unbroken flow state in which you freely and happily push the boundaries of the universe. If however, you are a person who wants to do research, but keeps finding other things to do instead (problem sets, surfing the Internet, going to buy your 3rd coffee of the day), you will have to train your own mind.

The two most important things I did to train my mind were: (1) read Virginia Valian's essay, *Learning to Work*, and (2) start recording my hours. You can also Google "Pomodoro method" or "Getting Things Done." Whichever method you choose, realize that it takes a time investment – actual training – to be a competent scientist.

Social (Dis)comfort

When talking to people on airplanes (a frequent occurrence), they often ask me what it is like to be a woman in physics. To which I say, "HAHAHAHAHAHA!" Actually, I usually manage to suppress that reaction and say something more diplomatic, which I will also attempt to do here.

On the plus side: I relate very well to other physicists. We are a tribe. I share the same precise, analytical, conceptually-oriented thought patterns that we all have been trained to embrace. I interrupt other people to say things like, "Actually, you are *wrong*, because…" on a regular basis.

On the other hand, I often wonder about the effects of spending the last 18 years of my life in a social environment in which I was profoundly uncomfortable. Being the only woman in the room can take its toll. At the beginning, I was worried that no one would take me seriously. I attempted to look as much like a guy as possible to fit in. Four years ago, I moved to L.A., an image-obsessed locale in which continuing this strategy would have amounted to social suicide. I now wear makeup - and occasionally a skirt - to work. It still makes me nervous. Fortunately, I now have enough lines on my CV that I figure people have to take me seriously. *Don't believe a blonde woman can do physics? Check my webpage, buddy!*

Fortunately, I do know a few other women physicists and engineers. Thank goodness. Even having one other woman around within eyeshot makes a big difference in my comfort level.

Taking Care of Yourself

Whenever any of my friends runs down a long list of problems that are making them stressed out, I tell them to go running. It is amazing how well this works – for me, anyway. Just give me 30 minutes of hard, aerobic exercise, and all my unsolvable problems magically sort themselves out into a manageable action list, with a set of clear priorities. (In times of extreme crisis, this may require more than 30 minutes.)

Despite my early history as the least athletic kid in an extremely athletic family (my dad was running marathons with my sisters when they were in high school), I have been an avid exerciser since the middle of grad school. It is one of the key things I count on to keep me sane – along with regular sleep, healthy eating, and time for rest and relaxation. I have many friends and colleagues who pride themselves on sacrificing it all in the name of science, staying up until 4 in the morning, and getting by on caffeine and Snickers. Hey, go ahead if this works for you… but I find I actually think better and produce more when I take care of myself physically and mentally. I suspect you will find the same.

I strongly value self-development, challenge, and growth. In my job and in my personal life, I devote considerable time and effort both to promoting my own development and supporting that of my students, colleagues, and friends. I believe in challenging yourself – but not to the point of pain. Think of it like cross-country running, or indeed any other sport: you want to train to meet your goals, to push yourself farther and faster, but not to the point of injury or debilitating fatigue. Academia is no different.

A final word for players of team sports: team sports are cool. You learn to cooperate with other people to achieve a collaborative goal. Your teammates help you; you help them. Unfortunately, science is not always this way. While cooperation and collaboration is essential, keep in mind that you will be evaluated on your own achievements. Always make sure that your contributions to a project are identifiable, and that you can point to what you did to contribute to its success. If you are extremely helpful and collaborative by nature, you may have to learn to be a little bit greedy: first-author papers really do count a lot more than ones where you are further down in the author list.

And if you are an egotistical, selfish narcissist? Taking up basketball or volleyball may be good for you.

Concluding Remarks

All right! That's my wisdom. Now go forth. Decide what you want to do, and do it!

Science is fun, thinking is cool, and learning something no one has realized before is awesome. I haven't told you much about my research in this essay, but it is really, really interesting (you can always take a look at my web page). I am happy and grateful to have a job where I can decide what I think is important, and then go and do it. It is a special privilege to have this freedom. I wish it for you, as well.

DIANE (BETSY) PUGEL

Bona Fides

Dr. Diane (Betsy) Pugel graduated with a Bachelor of Science degree in Physics from the University of Michigan – Ann Arbor, where she was a National Science Foundation Scholar. She received her Ph.D. in experimental condensed matter physics from the University of Maryland – College Park. She didn't stray far from University of Maryland, as she now works at the NASA Goddard Space Flight Center in Greenbelt, Maryland. In her role, she pursues innovation in a wide variety of applications and searches far and wide to work on interesting challenges for the Agency's missions. She has had the privilege of working on the Space Shuttle, the Stardust Sample Return Capsule, and the Orion Crew Exploration Vehicle; she is currently working on a mission to do science on Mars and a mission to study Jupiter's icy moons. She is committed to sharing her love of science and innovation with the public through hosting science radio programs (she has even done this at the Burning Man Festival) and through programs at local schools, the Smithsonian Institution, and the U.S. Science and Engineering Festival. She has received many awards from NASA and from the community at large.

A Link Between Art and Science

I thought that my life would take me to a world of poetry and art in Greenwich Village, in New York City. I was in middle school, impressed by the fluid feeling that art and literature had allowed me to express and hardly impressed by the math and science environment around me. There was nothing worse to me than to be in math class. In geometry class, I was the butt of jokes – a girl who seemed to do quite OK in class, but the boys sang this song to me, in true 80s style, to the tune "Maniac:" "She's a brainiac, brainiac in math class." In science, I had a teacher who had a piranha, Oscar, whose tank was in the back of the classroom. If paramilitary men had their bulldogs, Oscar was my middle school science teacher's equivalent of their bully mascot. Oscar had nothing to do with science and everything to do with classroom control. He hardly brought out my curiosity for the subject and did more to fuel my disdain for my teachers.

I was a good student in middle school, yet I was yearning for a deeper intellectual challenge and was worried about high school. I saw my two older brothers and older sister have the same deep intellectual yearnings and the challenges, frustrated by the social pressures that our local high school had to offer. When I was in eighth grade, I knew I had to make a dramatic change. I told my parents that I wanted to go to an all-girls high school. I felt that there, I could be away from the boys who teased me and focus in on my intellectual pursuits without having to put up with the raging hormone soup that my older siblings had faced. My parents found the money, which was hard in a household with five children, and I went to a private high school.

In high school, my freshman science teacher saw the hidden talent in me to do science. She dug it out of me, since any feeling that I had towards the subject went totally dormant in my middle school days of being a "brainiac" and at the mercy of a bully piranha man. She knew that I was comfortable challenging authority and made it a point to create situations where I could be challenged in an environment of growth. It was from her that I first heard about gender and science. She used to say: "Pugel, if you are going to be a physicist with that

214

blond hair of yours, you need to know your stuff!" So, a few days later, in protest, I dyed it black. Thankfully, the dye was temporary, fading my hair to green temporarily, but I learned a valuable lesson with my black hair – people really did think that if you were a woman and interested in science, you'd be taken more seriously with dark hair. No kidding. In my high school environment, in policy debate, my judges had different responses to comments that I had made in rebuttals during a debate match when I had black versus blonde hair. So, I knew I had challenges ahead and thankfully was learning the tools to help me through this process!

My teacher introduced me to the link between science and art by pointing me towards biographies of the more "mystical" physicists, like Einstein, who saw the deeper mysteries in physics as something of beauty instead of something as strictly sterile or logical. That was music to me. I started to read about scientists who also did art and artists who took advantage of the science that they understood to create great art. I started to read about special and general relativity as well. I was smitten with physics.

My road in undergrad and grad school was filled with many different kinds of learning experiences – some I sought out and some just happened to me. I did a lot of research as an undergrad, seeking out different kinds of physics to taste and see what I liked. In undergrad, I worked in experimental high-energy physics, experimental condensed matter, theoretical cosmology, and general relativity. Each time, I sought out a professor or an area that I wanted to learn and took the initiative to talk to them. In part, I picked the subject I was interested in learning more about and in part, I picked the person or research group. It was clear to me that the local culture mattered just as much as the work that I was doing.

In my first year of graduate school, I wrote an essay that eventually was published in the American Physical Society news that discussed how women can become derailed at different stages of life as they pursue careers in physics. Every time I read that essay, I think back to my graduate school experience and wonder if I had a sixth sense of what was to come for me. Graduate school was challenging in so many ways, more than I can express in the space allotted for this essay. However, even in the midst of these challenges, my love of physics and support from family and colleagues helped me get to where I am today.

At NASA, I have found that I have so many options – NASA is a big place! In this big place, I have been fortunate in being able to work on projects that cater to my curiosity and give me a venue to fully express myself as a scientist and manager. I have worked underneath the Space Shuttle Discovery, led the charge for the manufacturing of the heat shield for the vehicle that will replace the Shuttle, and worked on astrophysics telescopes and missions to Mars, the moon and the Outer Planets. Each day is a blessing and a challenge!

Reflections and Recommendations

Take the time to explore your likes and dislikes. It is really OK to try something out for a while and then decide it's not for you. Perhaps later on, you'll revisit those likes and dislikes, so do not be afraid to try something again later in life!

Balance is important for me, but not for everyone! Some folks thrive on the full immersion that a career in science and engineering brings. I have learned that my thinking process is somewhat unconscious, so I incorporate a lot of non-science activities into my life and have lots of non-science friends to help that process along. For me, that is a true need! Without it, the gears in my head feel all gummed up.

Do not be afraid to fight. I have played soccer and rugby, and in those sports, I have learned when to fight and how to fight. Girls are often taught not to fight. Sports are a great venue to learn how to be a team, to learn to fight for yourself or your team, and to learn when you're playing a game where the referee is just a bad referee and the situation will be out of your hands (this round!). Sports are a great metaphor for me.

Different is good. Different might be hard sometimes, but different is good.

JUANA RUDATI

Bona Fides

Dr. Juana Rudati earned her Ph.D. in experimental atomic and optics physics at State University of New York at Stony Brook. She worked exclusively with visible light lasers during her graduate studies. During her employment as a postdoctoral researcher at Argonne National Laboratory (ANL), she was stationed at Stanford Linear Accelerator Center (SLAC). There she expanded her research experience to include X-rays, studying laser-pumped/x-ray-probed ultrafast processes. As a Project Manager at Xradia Inc., Juana's work currently concentrates on research and development of high-resolution x-ray microscopes and their applications. She has led multi-million dollar advanced research projects in x-ray technology, federally funded by DARPA and ATP. Juana is actively engaged in the x-ray microscopy and x-ray optics community presenting at scientific meetings, publishing scientific papers and collaborating with academic researchers. She actively participates in efforts to promote women and minorities in science through participation and presentations at meetings and locally by mentoring junior female researchers. She lives in Orinda, California, together with her husband (who is also a physicist) and their 1-year-old son.

In the U.S. at 14

High school is a blur to me. I emigrated with my mother from

218

Argentina to Florida in the middle of my sophomore year. I tried my best to cope with the handicap of having to use English at school, which was a second language to me and not used in our home. Neither my mother nor my father had university degrees. They had started but never finished, which is not uncommon in Argentina since university is free. This was very influential for me and probably the strongest driver for my determination to succeed in school and finish university. On the high school aptitude test I performed very well in math and poorly in English (I would have done better in Spanish!). The scores were combined, and my total score was not good enough to get me directly into a university. Instead, I went to a local community college.

The community college environment was a perfect fit for me, and my skills quickly blossomed. I took classes that were challenging, interesting, and affordable, and I had teachers who considered teaching their primary vocation. I went to California with the honors club and to several colleges with the math team (fortunately, all expenses paid). I told everyone, particularly my physics and math teachers that I wanted to go to a university, and they were very supportive. Looking back at this time, I would definitely recommend this path over going straight to a research-centered university, especially if money is tight.

Needless to say, I was well aware that I needed not just good grades, but excellent grades, and mostly from honors classes when they were available (which I think were actually easier). Without financial support from home, I was grateful that the college's job placement office helped me find a flexible job that allowed me to take classes in the morning, work in the afternoon, and do my homework in the evening. The afternoon job and Pell grants paid my way through community college, and my dream of attending university came true when my grades and math team championship got me into UC Berkeley!

Perseverance through Difficult Times

During my first year at Berkeley, I received a phone call from my mother – she had broken her arm. She was OK but in a lot of pain. They had to stabilize her arm with nails, and she had it in a cast. She was unable to perform her job as a nurse assistant for an extended time, and without medical and disability insurance, she was worried about how she was going to pay her living expenses and medical bills. She had no paid holidays, no sick days, no savings, and still had to support my high school-aged sister. I immediately sent her the amazing sum of money that I had been able to save from working while studying: $500, a fortune for me. I considered dropping out of school: "Just temporarily," I told my mother. I said, "Just until you can work again; I will work full-time and pay the bills." She had none of it. Most of my informal mentors at the time agreed with her and encouraged me to think long-term. I am grateful for it. Twenty years later, I have a Ph.D. in physics and have helped my mother pay off the $40K in credit card bills that she amassed during those months that she was injured. She is one of those un-insured people that eventually pay every one of their medical bills – but that is another story. I realize now, however, that sudden financial need is probably one of the most common reasons students end up dropping out of university.

Graduate School

Physics graduate school was much easier for me than my undergraduate studies. One big reason is that graduate students in the sciences usually get paid – not much, but enough to pay the bills. In addition to my graduate stipend, my salary was supplemented by a scholarship and the Alliance for Graduate Education and the Professoriate (AGEP) program, which also provided excellent support and advice from administrators.

There are people who leave graduate school after the first two years with a master's degree, but I think that's when the fun is just beginning: when you finally stop simply learning physics and start to explore and explain it. You get to measure and experiment with things that very few people (possibly none!) have dealt with before.

I believe some students may get discouraged looking for a good thesis advisor. I worked really hard to find a suitable match for me. The advisors that had a good reputation had way too many students and – in my view – the others ... well, they were not a good match. Thankfully, I was allowed to be advised by scientists at the nearby Brookhaven National Laboratory (BNL), which increased the pool of possibilities. I found my match when a scientist from BNL came to discuss his research at a graduate seminar. After talking with him and reading his papers, I concluded that he was the right advisor for me. The only other thing I found stressful about these first few years in research was not knowing if I would actually be able to do work worthy of getting a Ph.D. Luckily, I've learned, just as many of us learn to ride a bicycle, the majority of students who work hard get their degree.

Conclusions and Recommendations

Be determined and stay in school. I wouldn't earn half of what I earn today if I did not have a Ph.D. in physics. If you don't have wealthy parents who will support you for your whole life, or if you want to be independent, choose a field of study that will pay afterwards. It is well-known that fields that use mathematics pay well, since graduates from these disciplines have many transferrable skills as well as demonstrated perseverance and stamina.

When you are an undergraduate, graduate, or postdoc:

Look for champions on your side: the more respected, the better. Anyone that has your best interest in mind is great: a more advanced student, the head of diversity affairs, anyone. I had both student and faculty mentors at Berkeley. They helped me get a well-paid work-study job tutoring other students. Others advised me to do a summer Research Experience for Undergraduate (REU) program, an excellent opportunity to check out universities where you might want to go to graduate school.

In industry:

Negotiate the details BEFORE you start working: title, salary, signing bonus, office space, reporting requirements, who you will report to, who will be your colleagues, what you will be doing. Get it in writing. Attend every meeting that you are invited to and make sure that you are invited to every meeting you should be in.

When you get to where I am:

It takes conviction, talent, leadership and many other desirable qualifications for a woman to be successful in a male-dominated field. When you are going against the norm, some people do not give you the benefit of the doubt the way they would for others. Take, for example, the time at UC Berkeley when I received a lower grade than I expected in one of those mega-sized physics classes. When I asked why I had not received full credit for my correct answers, the grader said that I had not shown my work. He could not understand how I had solved a simple single-variable equation, and he said I had to show EVERY step. He suspected that I had copied from someone else. At that point, having just competed in several mathematics tournaments, winning twice in the state of Florida and once getting the fourth highest individual score, I didn't see any missing steps. I had trained for that! He was un-convinced. The lower score and the anecdote remained, but as just a minor bump on the road to success.

After my postdoc years, I did not get the National Laboratory job that I wanted. Instead, I found a job at a company, doing research, at a faster pace. I love the feeling of being in a small start-up venture. Everyone is on the same side, doing their best, pulling together to make the company successful and the product the undisputed best. It has been very exciting and better remunerated, which brings me to my last point of why one should study physics: many of us are not doing what we set out to do when we first decided to study physics (you will find that many people wanted to be professors). But looking objectively at our current situation, we are in a similar, sometimes even better, occupation. We do a job we can be proud of. We contribute to the betterment of our society by adding to the body of knowledge. And most of us (if not all) don't hate Mondays and aren't counting the hours until Friday afternoon.

MICHELLE D. SHINN

Bona Fides

Dr. Michelle D. Shinn was born and raised in Oklahoma and received her Physics degrees at Oklahoma State University (OSU). After receiving her Ph.D., she joined Lawrence Livermore National Lab (LLNL) in 1984, working in the Laser (Y) Division, until 1990, when she left to join the faculty at Bryn Mawr College as an Associate Professor of Physics. In 1995, she joined Jefferson Lab as a staff scientist. From 1996-1999, Michelle led the design, procurement, and installation activities for the IR Demo free-electron laser (FEL) optical cavity, transport and diagnostics, and from 1999 to 2006 performed the same duties on the Upgrade FEL, the world's highest power tunable ultrafast laser. She was then promoted to Senior Staff Scientist with the title Chief Optical Scientist. In this role she sets the specifications of the FEL optical systems and analyzes their performance. She actively collaborates with a number of teams that use the FEL, and in particular, pursues her own research on the characterization of dielectric thin films for laser applications and laser-induced damage of optical components. In 2012 she was elected to Fellowship in the American Physical Society. She is a frequent public speaker on the subjects of dark matter and energy. Michelle enjoys amateur astronomy, beekeeping, and camping.

Beginnings: Asking Why?

I reached out and, righting the toy top, grasped the ball atop the spindle and began pumping it to set it spinning rapidly. I released it, and it spun away from me, brightly flashing. As it slowed, the top began to lean over and rotate about the vertical. This precession rate grew faster and faster until the top finally fell over and skittered away until it came to a stop against the drier. I looked up at my mother, who was talking to my grandmother while doing the ironing, and asked, "Why mommy? Why does it do that?" My mother, interrupted, looked down at me, then at the top, and said, "Because." I grew upset, "No, because! WHY?"

My life is one spent searching for the answer to "why?" I am fortunate that all the defining moments of my life are etched in my memory: by age 7 or 8 I was receiving chemistry and geology "kits" for Christmas, massive metal-cased containers that demonstrated the wonders of science; by 12 I built a cloud chamber, with the help of the book The Scientific American Book of Projects for the Amateur Scientist, borrowed from the local library. I knew by the eighth grade that I wanted to be a physicist, a desire crystallized by a simple experiment. Our earth sciences teacher, a balding gentleman with nicotine-stained fingers, had us hold hands and stand in a circle. At one point, the circle was broken and the students at each end held a wire that was connected to a World War II field telephone. As the teacher slowly turned the crank that would have normally rung the bell at the receiving end of the phone, he was actually running a generator. As the current went through the circle of students, our wrists twitched. A dramatic – but not too uncomfortable – demonstration of Lenz's Law and electrical conduction. After the experiment, I asked, "What science describes this effect?" When he told me the closest science to this and the other demonstrations he'd done was physics, my mind was made up. I would study physics when I went to college. I believed then, and still do, that physics underpins all the other sciences.

Answering the Questions – Becoming a Physicist

Fast forward to the fall semester of my senior year at university. The semester was just starting, and I was on the fence as to what to do with my physics major. Try to get a job teaching high school, or go to industry? Graduate school? I was feeling a bit drained after the rigors of junior year, the wash-out year for physics majors at Oklahoma State. Physics is a hard subject for most people, but I was so compelled to be a physicist, to satisfy my questioning mind, that I was going to graduate with a physics degree come Hell or high water! What spurred me on were the flashes of insight that came more frequently – *that is so heady!* – and, I suppose, because I am not a quitter. Once I have a goal, I achieve it, which is a good quality to have in physics. Looking back over that time and the years gone by, I like to say that rarely does mastering a talent come easily. Look at stories about how Olympic athletes train, and read the biographies of great scientists and other famous people who are so good at what they do. They struggled to achieve the feats they are known for! Perhaps you will be one of the few for whom physics comes easily. But if you are like most of us, do not be discouraged if you find physics to be challenging.

So I chose to go onto graduate school, and I received my Ph.D. After graduation, I did not follow the traditional path of most of OSU's newly-minted Ph.Ds. I took a position as a staff physicist at a Department of Energy (DOE) national laboratory, LLNL. I had also considered industrial positions, an act that my peers considered "prostitution." A harsh indictment, but one has to remember that all my peers in graduate school had no clear concept of an industrial physicist's career. The consensus was that the only reason anyone would "go to industry" and relinquish academic freedom was to have a higher salary. I wanted to do interesting condensed matter physics, particularly on new solid-state laser materials, and the institutions with the best facilities – and the majority of the best people – were either in industry (back then, TRW and Hughes Research) or at the national labs, especially LLNL. I also thought, and still think, that the concept of academic freedom is invoked too frequently by those who don't understand the limits imposed by the funding agencies. They can't fund every good or interesting idea one might have, so the majority of scientists in academia, particularly those starting out, are not free to pursue all their research interests. Finally, I had a newborn daughter and felt that my spouse and I needed as much income as possible to give her a good upbringing, and it was true then, as it is now, that the starting salary of a post-doc or an assistant professor is less than that of a freshly-minted Ph.D. working at a government or industrial facility.

It was 1983 when I graduated and not the best of economic times. I interviewed for several positions and chose to work at LLNL. It was a cultural shock, moving from a small intimate group of graduate students and professors in Oklahoma, to Northern California, where I was the only female Ph.D. out of 90 in the inertial confinement fusion division. It was exhilarating, particularly as I had to create a new spectroscopy lab and dive into data collection on the samples that were coming in. It was also my first time to participate in group meetings of about 10 folks, where once again, I was the only woman. I learned that I frequently seemed to not have a voice. Listening to our group leader spell out a problem, I would raise my hand and make a suggestion and get … no response. Nothing. And then, within 10 minutes a colleague would make the same comment and be praised for it! It was so frustrating! After considering my options, which included complaining to Human Resources, I decided that the best way to deal with this was, at meeting's end, to go to my office and write a memo where I would spell out what I had said and explain why I thought it was a good solution. Did it make a difference in his behavior? Initially no, but over time, and with a growing list of accomplishments to my credit, he and my colleagues did listen to me. Talking to my younger female colleagues about their experiences, the same "deafness" still occurs to varying degrees amongst some of their male colleagues. My recommendation is to follow my example and document what you said. And back it up with action: produce results!

I could write a chapter, and someday will, about my time in academia. As it turned out, it wasn't a good fit for me, so I returned to the environment of the national labs but with a commitment to do research with students. The one maxim I created as a result of being mostly on my own, research-wise, is "Not making a decision is a decision to fail." In other words, while it is good to do some reflection, research, or planning before making a decision, don't delay it too long out of fear it will be wrong. Will you sometimes make a wrong decision, or come to a wrong conclusion? Sure! But, as long as you can say to yourself, if no one else, "Here are the logical steps I took to come to this decision or conclusion," then at least you can determine what went awry and be the wiser for it.

I have now been at the Thomas Jefferson National Accelerator Facility (Jefferson Lab, or JLab for short) for over 18 years. My Division's leaders have given me opportunities to work on some of the most amazing projects. My opinions are listened to – not always accepted – but acknowledged. In return, my team and I have made and continue to make advancements in the FEL state-of-the-art. Along the way, while operating the FELs for users, I've had some adventures: times in the Control Room where hardware acted up and we, as a team, pulled a rabbit out of the hat and saved the experiment. Not life or death situations, like in the movies, but exciting nonetheless!

Advice to the Next Generation

Looking back over a career that has spanned nearly 3 decades (admittedly it doesn't seem that long!), I'd like to conclude with some advice that might help make your professional journey smoother. First, seek out mentors who can offer you advice and support by answering both specific questions and providing general guidance. They will help you make contacts, suggest opportunities to follow, and share anecdotes of their experiences at a similar stage of their career. I found that their stories lifted my spirits when I was down, when I felt inadequate, and when I was challenged by the coursework. Your mentors will also push you to aim higher when they believe you aren't working to your potential. Great mentors don't necessarily have to be other women, or even other physicists. When I entered university there were no good female mentors, but I had several when I was at LLNL. At Bryn Mawr there were a number of new female hires, and we had each other to consult and commiserate with. At present, most of my mentors are men, and at this stage of my career that's fine. With the increasing participation of women in physics, I think the next generation will have an easier time finding female mentors if they so choose.

Second, as a budding physicist just starting your studies, I suggest you read broadly in science and technology. When in university, attend seminars and colloquia. Early on, you will get lost after the first few slides, but you will see the way the speakers do physics, and what it is that excites them. Who knows, perhaps you will find a subfield of physics to follow.

I can think of no other career I would rather have. Physics enriches your soul as much as it enriches your mind. Each day provides a new opportunity to learn something, and I am never bored. To paraphrase Francis Bacon, the universe is your province, with rich and varied questions to be answered in every subfield, by both experimentalists and theorists. I am excited for you, just starting your career: welcome to the party!

ELIZABETH H. SIMMONS

Bona Fides

After earning an A.B. in Physics magna cum laude from Harvard University, Dr. Elizabeth H. Simmons spent a year earning an M.Phil. in theoretical condensed matter physics at Cambridge University as a Churchill Scholar. She returned to Harvard for her doctorate in theoretical particle physics and stayed on for a postdoctoral fellowship. She was hired as a physics faculty member at Boston University and moved up through the ranks to become a tenured associate professor and an associate department chair. Elizabeth is presently a Professor in the Department of Physics and Astronomy at Michigan State University and Dean of Lyman Briggs College, a residential undergraduate college focused on science in societal context. She is a Fellow of the American Physical Society and the American Association for the Advancement of Science. Her research focuses on the origins of the masses of the elementary subatomic particles - particularly the W and Z bosons that transmit the weak nuclear force and the heaviest known particle, the top quark. She enjoys teaching physics courses at all levels, from introductory freshmen courses to graduate classes. A central part of her mission as an educator is encouraging more students (especially those from groups now under-represented in physics) to consider studies and careers in the physical sciences.

Becoming a Physicist

I have always loved nature and mathematics. As a child, I spent a lot of time outdoors catching snakes and turtles (my parents accepted all sorts of odd pets) and reading nature magazines about exotic creatures from other parts of the world. By the time I was in high school, I had been allowed to skip ahead in mathematics and language classes, which encouraged me to do a lot of learning on my own. So when it came time to pick a likely career path, I decided to focus on the physical sciences where I could apply mathematics to the natural world and where it seemed there would always be a lot to learn.

Several experiences steered me towards physics in particular. My first-year physics teacher in high school was a lively instructor who insisted that we think on the fly, analyze tabletop demonstrations, and determine whether our answers made sense. His class was challenging and exhilarating. I was also lucky enough to have a mentor at Bell Laboratories, where I was a member of a Boy Scout Explorers troupe (this part of the Scouts was co-ed). My Bell Labs mentor encouraged me to attend a Summer Science Program (SSP) in astronomy and later took me on as a research student for my senior year of high school and first couple years of college.

SSP provided a first glimpse of how all-encompassing life as a scientist could be. I spent six weeks as a member of a team utterly focused on learning the advanced mathematics and celestial mechanics needed to determine the orbit of an asteroid while simultaneously making the midnight telescopic observations required as input for the calculation. Because I came from a small school, this was my first experience among a full three dozen peers with a similar background and interest in science. I came back certain that I wanted to be a physicist – and then spent my senior year of high school working for my Bell Labs mentor doing research on experimental condensed matter physics with astrophysical applications. We used a particle accelerator to shoot ion beams at frozen molecular gases (ices) in order to deduce whether frozen water could survive in shadowy craters on the moon or how erosion of ice grains in space might give rise to clouds of organic compounds. Being accepted as a full-fledged member of the research team able to run the accelerator, design experiments, and serve as a co-author on published journal articles was an incredible experience.

During college, I obtained research experiences in different areas, from electronics to biophysics, while taking all the advanced courses in physics and applied mathematics that fit in my schedule. When I was offered a one-year Churchill Scholarship to Cambridge University in England, I decided to do a master's degree in theoretical condensed matter physics at the Cavendish Laboratory because I felt best prepared in that area. Working with my Cavendish mentor and a fellow graduate student on crystal symmetries and the properties of spin systems was fascinating. Again, there was the constant drive to learn about new ideas that went beyond course material. While engaged in that research degree, I attended a lecture class in quantum field theory and particle physics in my spare time; the sheer mathematical beauty of the subject convinced me to try particle physics for my doctoral studies when I returned to the United States. And that's the field where my research has focused ever since.

After my doctorate, I was fortunate enough to be hired immediately as a postdoctoral research fellow and then to find a faculty job after just one fellowship. Obtaining tenured (permanent) status five years later was a huge relief; now I knew that I could be a professional physicist for the rest of my life.

Becoming a Scientist

Being a scientist, however, means more than knowing how to solve equations and write papers. You need to understand what kinds of questions science can address, how societal constraints have influenced the course of science over time, and how science can give back to society.

Part of what has made me a capable scientist is the fact that there has always been more in my life than science. I have been a tutor and teacher since I entered college, both in English and in math and science. I have been an amateur musician since childhood, in voice, recorder, and piano. I take time out for sports like fencing (yes, swordfighting), squash, biking, and hiking in the mountains. Due to the advent of smartphones, I always have several books in my pocket – mysteries, science fiction, biography, and social history. When traveling for research, I take time out to visit museums and national monuments in countries around the world. And I still love observing wildlife, even in my own backyard where the songbirds and squirrels are accompanied by deer, raccoons, rabbits, frogs, snakes – and snapping turtles the size of small watermelons.

My current role as Dean of a residential undergraduate science college brings me into close contact with scientists of all fields along with historians, philosophers, and sociologists who study what makes up the essence of science. Much of what I do relates to hiring faculty and guiding them along their own career paths to success as teachers and scholars. Working with people in such different fields has helped me become aware of the unstated assumptions physicists, like myself, use in our work; this has made me better at explaining the beauty and utility of physics to those outside the field. Likewise, I have had a chance to learn more about the key questions and methods that other scientists use to explore the natural world. I think that any student of science should take advantage of the opportunity that college life offers to learn how different fields fit together and how they historically developed from earlier approaches to studying nature.

Being a Woman in Science

As a woman physicist of my generation, I have always been aware that there are few of us. Fewer than 10% of the full professors of physics are women, after all. Whether as a student, a research fellow, or a faculty member, I have often been and still often am the only woman in the classroom, on the team, or in the meeting.

Unfortunately, because we are so much in the minority, most of the women physicists I know have shared stories with me about their experiences with gender-related discrimination and harassment; I have shared mine with them as well. Talking over our common experiences has helped us figure out ways to get past these trials and move forward to better situations and opportunities. It has also inspired many of us to advocate for a more equal role for women in physics, whether in national labs, universities, professional societies, or research institutes. We have worked to promote more women into leadership roles and have spent time talking with younger women scientists about how to navigate in the scientific workforce.

Fortunately, there are many wonderful male scientists who fully believe that scientific talent is not restricted to one gender and who work alongside us to try to change the culture of everyday scientific life. I am lucky enough to be married to one of them. We teach courses together, are co-authors on research papers and co-advisors of doctoral students, and we have spent the last two decades jointly managing a household and bringing up two children.

While not every woman scientist wants her life partner to also be in the same field, I think it is especially important for women scientists to choose partners who will encourage them to persist in pursuing their career goals. Because we are still likely to encounter doubt about our competence or dedication within the workplace, being able to rely on a supportive family can be crucial.

Parting Words

To the young women of future generations, I advise you to hold fast to your love of science. It is what will see you through those frustrating times when the vacuum seal won't hold, the code is full of bugs, or minus signs dance maddeningly through your calculations. Remember why you find science fulfilling – and recall how your special creativity or insight has enabled you to design the manifold or interpret the data in a way that makes a difference. This is what makes science a worthy career for you and makes you a worthy scientist.

I also counsel you to analyze your dreams of a scientific career so that you understand what it will take to get you there. Know what courses you will need to take to attain the next level of mastery. Learn how to choose a research advisor who will actively teach you the professional as well as scientific skills the job market requires. Discover what to ask for in your job negotiations to ensure that you have the money, time, and resources essential for success. How will you find out these things when you need to know them? By talking to other supportive scientists (women or men) who have been there before you. We are out there, and we want to help you. Just ask us.

ELMA BETH SNIPES

Bona Fides

Ms. Elma Beth Snipes received her Bachelor of Science degree in Materials Engineering from North Carolina State University (NCSU), where she worked for the Engineering Research Services Division. There, she was introduced to x-ray diffraction, which became a life-long love affair. While working at Babcock and Wilcox's Lynchburg Research Center, she pursued a Masters of Materials Science at the University of Virginia. She performed applied research on nuclear control rod poison, failure analysis of steam generator and reactor components, and x-ray diffraction texture analysis. After several all-expense-paid "vacations" to nuclear power plants, she left the nuclear field to pursue a career in x-ray diffraction. She landed at TEC, a small high-tech company that was developing portable x-ray diffraction systems for residual stress analysis. Beth has helped to develop three generations of x-ray diffraction systems for TEC. She holds a patent for MAX, a Miniature Advanced X-ray system. True to form, she continued her studies at the University of Tennessee (UT), where she has accumulated 32 hours toward a Ph.D. in Materials Engineering. The degree was put on hold while she pursued her career and motherhood. The degree is on her retirement bucket list.

Adventures with X-rays

Dreams and Goals

My parents grew up on farms during the depression. My dad served in World War II and came home to find employment difficult, as he only had an 8th-grade education. He earned his high school diploma from Campbell College and transferred to NCSU, where he studied agricultural education and graduated in three years. My grandmother died when my mother was a child, so in many ways my mother had to be an adult for most of her life. When she graduated from high school, she was able to get nurses' training and became a Licensed Practical Nurse (LPN).

My family and friends described me as a tomboy. I spent my formative years in Raleigh, NC playing basketball and cowboys and Indians with the neighborhood boys. The den mothers sent me home when my playmates had their weekly Boy Scouts meeting. I joined the Girl Scouts but was disappointed to find that the troop did not go on camping trips like our male counterparts.

My first loves were animals, reading, and music. Unfortunately, living in the suburbs limited my contact with animals. I developed a following of neighborhood dogs, but I never was allowed to have my own dog or cat. When I was not outside romping and running, I was reading. I loved classics, mysteries, biographies, and animal stories. My mother made it a habit to read daily to my sister and me, and I suppose I caught the bug. As for music, at the ripe age of 10, I begged my parents for a piano, which they sacrificed to provide – as long as I promised to keep it dusted. I practiced regularly, so eventually they released me from that promise. From an early age, these loves were shaping my life.

My neighbor drove an early '60s model Corvette convertible. I told my dad that I wanted one when I grew up. He told me I could have anything I wanted as long as I wanted it bad enough and was willing to work for it. These words had a profound effect on me. I started putting together my wish list for my life. At the top was a farm with horses. A Corvette was also on the list, but not much mattered after the top spot.

Growing up in the 1960s, I was expected to finish school and find a husband. Since my parents wanted only the best for me, "finish school" meant earning a college degree. In my rebellious teen years, I left home and school, but I quickly returned to school when I found how unforgiving the real world was. I petitioned the school board to allow me to graduate early, and they granted my request. I applied to college without knowing what I wanted to do when I grew up. I wanted to become a vet, but the vet schools would not take women into their large animal programs, so I settled for a pre-med path.

I worked at JC Penney's to help with college expenses. One day, a woman came in to inquire about fabric shears. I showed her the new lightweight, super-sharp Fiskar shears. She indicated that she was not familiar with these scissors and that she was a metallurgist. I asked her if she was at the university, since that's the only place I had ever heard of a metallurgist, and she said she was on staff there. When I told her I was a student, she asked me if I wanted a job. What an opportunity! I started immediately, switched majors to Materials Engineering, and was able to get a scholarship every year through the completion of my degree. Mea Fiedler discovered me and taught me x-ray diffraction. Mea and her boss, Dr. Hans Stadelmaier, provided the foundation that has shaped my entire career.

Career Development

Prior to my last semester of college, I married an engineer who was working for Babcock and Wilcox. I interviewed with their Lynchburg Research Center (LRC) and was offered a position after graduation. The human resources employee commented that I *also* had a great GPA - I guess I was naïve to think that I was hired on my outstanding academic record alone. At that point, however, I was simply relieved that we had cleared the first hurdle of being a professional couple gainfully employed in the same city. Later, my husband and I went on several job interviews together to get around the issue of a two-career couple. Interviewers often asked what my husband would do if I was offered a job. Although it was never explicitly stated, the implication was my work was never as important as my husband's. I chose not to believe this, but I had to be creative to keep my career a priority.

I enjoyed my work at LRC, where I worked hard and learned a lot. My boss pushed me to attend the local American Society for Metals (ASM) meetings. Meeting notices always encouraged members to bring their wives - even I was asked when I'd be bringing my wife! When I volunteered to help the executive committee, I was welcomed with open arms. As a result, meeting notices now suggest that members bring their spouses. This was a small victory, but it was progress. I continued to volunteer with ASM, and the networking and recognition have been a tremendous boost to my career. I encourage young people to make contributions to their technical societies - the benefits continue long after the efforts are made.

One interesting career experience was traveling to nuclear plants to oversee the extraction of failed components. In the 1970s, the plants I visited did not have women's change rooms (where anti-contamination suits were shed and scans were made to ensure radioactive contamination did not cross into the clean areas). At the end of my first trip into containment, many of the male workers followed me into the change room hoping to get a glimpse of female anatomy or underwear. Luckily, a woman colleague had suggested I wear a leotard under my clothes for modesty. They were either disappointed or relieved to see my "underwear" covered me from my neck to my ankles. I learned an important lesson – always do my homework and always maintain a sense of humor.

My daughter Mary was born in 1982. I never knew how much I could love a person until Mary was born. Around that same time, my husband was offered a position with a fast-growing small company in Knoxville, TN. I finally had the opportunity to live on a farm with real horses! Always looking for a way to meet my goals in different circumstances, I decided to leave my job, stay at home with my new baby, and, in my spare time, pursue a Ph.D.

After a couple of years, I longed to return to my career. TEC was looking for someone with x-ray diffraction and nuclear experience. Thank you, God. My salary more than doubled despite (or maybe because of) staying at home. On my first day, I went into the office, signed some insurance papers, and boarded a plane for Philadelphia. I spent the weekend learning how to make residual stress measurements with a portable diffractometer. Through the years at TEC, I've been able to apply my knowledge of x-ray systems towards developing small, portable diffractometers that can be hand-carried into the field and can make measurements in minutes.

When my daughter was around five, I brought her to work so she could see what her mother did during the day. I was so proud of her when she called her best friend to tell her she wanted to do what her mommy did when she grew up. Fascinated by the conversation, I listened as Mary continued to tell her friend that I had a machine at work that gave candy when you put money in it. It's interesting what children value. Mary has since gone on to obtain her degree in Aerospace Engineering, and she is on the fast track at an international oil exploration company.

My son Phillip, of whom I am equally proud, is a senior studying Electrical Engineering at UT. Not only did he follow his parents and sister into the sciences, but he is an excellent musician. He has found it odd that some of his female classmates are more interested in marriage than pursuing careers. He encourages them to become a productive member of society through education.

I have been able to have it all, but it took extra work on my part. My pregnancies restricted my career because of potential harm to my babies from radiation. I had to be creative to continue my work and protect my children. I spent more time reading, researching, and delegating - developing new talents that would help me to advance. When my children were sick, it was my responsibility to take care of them. Luckily, I was in a position to work from home on the days that Mary or Phillip were unable to attend school. Life isn't always fair, but I have found that if I continued to show value to my employer, I would be allowed to work around inconveniences.

Lessons Learned

Nothing beats knowledge of your field coupled with persistence. You can have anything you want if you believe it's attainable, and if you're willing to work hard enough. Sometimes working hard translates to working smart. Sometimes, you have to have a backup plan for the times that things don't go exactly as planned. I didn't start out thinking I'd be a scientist—I just knew that I loved learning.

Some people will try to put you into boxes so they can feel comfortable with who they are. Some of my fellow classmates were convinced that women were not as good at math. Some professors suggested that I learn to type at the local women's college, and stay there. I've been told that I would not be promoted or offered a job simply because I was a woman. Yes, it hurt my feelings, but it also gave me the kick in the pants to pursue my goals via a different route. Without the support of my parents and mentors, I probably would have convinced myself that I was not capable of learning science and engineering. But as life has progressed, I have learned that not only could I excel in a non-traditional field, but most people with a curious nature can too.

Having a good sense of humor and believing in God have been key to my staying on track. Of course, God also has a good sense of humor. Why else would He have allowed me to have a farm with horses, cows, and chickens? I didn't realize when I dreamed of farming that simple acts such as a few cows getting out of the pasture would result in an entire day spent building a new gate, replacing batteries in farm equipment, and re-drilling holes so the gate would be at the proper height.

My core belief is that we were put on this planet to make a difference and to give back by developing our talents. Although there were plenty of people who wanted to see me fail, there were those that took the time to encourage me and to help me to believe in myself. My heartfelt thanks go to Royce and Elma Snipes (parents), Mea Fiedler, Inge Simonsen and Dr. Stadelmaier (teachers and mentors), Joan Hutchison (piano teacher and friend), Mrs. Johnson and Mrs. Thompson (librarians that supported my love of books); and Mary Olsen and Phillip Pardue (children). Yes, there are many others, and I'll get in trouble by not mentioning them. My point here is that we need to encourage each other and be thankful for those wonderful people who help to make our lives better. I have found that if you seek out other successful people, respect their time, and take their advice, you may avoid some mistakes. Of course, some mistakes can also teach valuable lessons.

I've always been a goal setter. I feel strongly that you may not get somewhere if you don't know where you are going. I urge you to dream big dreams, and don't let others take your dreams away. The job in an x-ray diffraction lab has turned into a career that is fun, fulfilling, and challenging. That Corvette that I wanted as a young girl has turned into four in my driveway. The latest one has a 6-speed manual transmission along with a G-meter — it gives fun with physics a whole new meaning. The console piano that seldom got dusted is now a concert grand that gives me an escape from the daily challenges of life. The dream of a horse on my own farm has become several horses along with cattle, chickens, dogs and cats, and a new learning experience. The love of learning and courage of taking a unique path has made a difference in my life. Trust me that it can make a difference in yours, too.

MEG URRY

Bona Fides

Dr. Meg Urry graduated summa cum laude from Tufts University with a Bachelor of Science degree in Physics and Mathematics. She received her M.S. and Ph.D. in Physics from Johns Hopkins University, completed postdoctoral fellowships at the Massachusetts Institute of Technology and NASA's Space Telescope Science Institute, where she became a tenured Astronomer and served as Head of the Science Program Selection Office before moving to Yale. Currently, Meg is the Chair of the Department of Physics and the Israel Munson Professor of Physics and Astronomy at Yale University, as well as the Director of the Yale Center for Astronomy and Astrophysics. She is the first female tenured faculty member in the history of the Yale Physics Department and is a Fellow of the American Academy of Arts and Sciences, the American Physical Society, and American Women in Science. Meg has received several prestigious awards and honors throughout her successful career. She is a leading advocate for increasing the participation of women in science, which she refers to as her "second career."

The Early Years

My profile isn't typical for physicists. I liked science but not more than any other subject – I liked everything. I was smart but not a classic geek. I loved math, which is very important for doing physics, though that wasn't part of the calculation. I never liked science fiction, which many of my colleagues (and my husband and kids!) read. I did read like a fiend, and I loved writing; I think this has helped me in writing scientific papers and in communicating science to my colleagues and to the public.

My parents – a zoologist and a chemist – really prepared me to be a natural scientist. Growing up in the Midwest and then on the East Coast, my sisters and brother and I naturally absorbed my parents' logical, methodical way of thinking. I thought everyone thought that way! I thought it was normal to ask: What do we know? What are the options? What further information do we need to find out in order to figure out the problem?

For example, on long family road trips to California, it was part of the drill to observe the natural world. After a picnic lunch, my mother would poke around in any available stream, turning over rocks and looking for worms, which she had studied in college. That was normal for us. It was part of looking at the world around us and trying to figure it out.

When I was quite young, in 3rd grade or so, I read a lot of biographies, including some of famous women - doctors, scientists, and pioneers, like Jane Addams, a political crusader for poor people, Amelia Earhart, the heroic pilot who broke gender barriers, and Victoria Woodhull, a leader of the American women's suffrage movement. Some of the biographies introduced me to scientists, including Elizabeth Blackwell, the first woman to graduate from a U.S. medical school, Clara Barton, founder of the American Red Cross and crusader for women's rights, Marie Curie, the discoverer of radioactivity and recipient of two Nobel prizes, and Maria Mitchell, a pioneering U.S. astronomer. These women were incredibly inspiring, and I yearned to do something significant, something pioneering, as they had done. But I still didn't imagine I'd do that something in science.

I credit my chemistry teacher, Miss Helen Crawley, at Winchester High School in Massachusetts, for getting me excited about chemistry. Before that, science was probably the least favorite of my classes. Later, as I started college, my parents, and particularly my father, himself a professor of chemistry, were extremely influential, suggesting that I take physics – even suggesting that astrophysics was an interesting field – and always, always encouraging and supporting me.

Astronomical Inspiration

I turned to astronomy much later than many colleagues, who as children were amazed by the cosmos. I was more likely to have had my head in a book than wonder where stars came from. The turning point was a summer internship at the National Radio Astronomy Observatory in Charlottesville, Virginia, the summer after my junior year in college. There I learned how interesting and how much fun astronomical research can be. The scientists were fun, too. They made science seem less solitary and more friendly – something to be done with colleagues, something to talk and argue about.

My senior year in college, I applied to graduate schools in astronomy and physics and ended up going to the Johns Hopkins University Department of Physics and Astronomy (then it was the Department of Physics). In the summer before graduate school, I worked with an X-ray astronomy group at the Center for Astrophysics at Harvard. More interesting science, more fun! This confirmed that when I would go to Johns Hopkins, I should look seriously into doing astronomy. I later landed a summer job at the nearby Goddard Space Flight Center working with their X-ray astronomy group. This led directly to my thesis research on blazars, an unusual kind of galaxy characterized by a relativistic jet pointed directly at us. (Blazars pointing elsewhere are identified as radio galaxies.) The jet appears brightened by many orders of magnitude thanks to an aberration predicted by Special Relativity. The notion that I could figure out what was happening billions of light-years away in the cosmos, from just a few particles of light gathered by our detectors – well, this was too cool to be believed. The combined scientific excellence and low-key friendliness of the Goddard high-energy astrophysics group is probably the reason I loved graduate school and that the stresses and pressures never seemed too much to bear.

Trials of a Trailblazer

When I started college in 1973, women's liberation was under way. Women were admitted to the elite colleges, federal statutes outlawing gender discrimination were passed, women's roles broadened; we all thought it was just a matter of time until women caught up with men, until the numbers of men and women in public and professional life were equal. Although there were stories of women being excluded from "male" spheres, those seemed like anachronisms. Egregious discrimination, like that endured by Barbara McClintock (according to Evelyn Fox Keller's 1983 biography, A Feeling for the Organism), was uncommon, or at least, buried behind closed doors. Sadly, more subtle forms of discrimination have taken much longer to diminish.

From college on, being a woman in science was rare, and I remember being identified by my gender. One professor used to address the graduate quantum mechanics class as "gentlemen and Meg." I once found pictures of naked men on my desk in the graduate student office. Later, as a young faculty member, when I told my graduate thesis adviser I was pregnant, he responded not with the traditional "Congratulations!" but with saying, "So, you want to have it all!" I smiled at the time but later wondered, why is it "all" for me and "normal" for you?

As a postdoctoral researcher at MIT in the mid-1980s, I was for a time the only female postdoc in the Center for Space Research. I became good friends with a woman graduate student in my astrophysics group and a visiting woman

247

postdoc in a space physics group – friendships that are still important to me today. My postdoctoral advisor, in fact, had lots of women in his group, which was rare elsewhere in the department. This ought to have made me realize that the lack of women had more to do with a professor's attitude than with the lack of talented women scientists. I sometimes felt as if I were on the wrong side of the gender divide, surrounded completely by male professors, some of whom dated or married their students or secretaries. There were acceptable roles for women, it seemed, but not as scientists.

One time a colloquium speaker in the MIT astrophysics seminar illustrated his talk about the importance of high spatial resolution in optical imaging with a badly out-of-focus slide. At his request, the colloquium host gradually adjusted the lens, revealing a topless woman in a grass skirt on a Hawaiian beach. Some (male) students laughed. The one other woman in the room and I were stunned and appalled. She walked out. I debated whether to say something but thought I'd be accused of lacking a sense of humor; I left after 20 minutes, having realized I hadn't heard another word the speaker had said. Later, at a party, I asked the colloquium host why he hadn't tried to forestall this "joke." He said the speaker was a guest, and it felt rude to chastise him. It definitely felt rude to be made to feel totally out of place.

Meanwhile, I was being told that women actually had an easier path than men, that universities were eager to hire women, that I'd be inundated with offers. This didn't happen. One young faculty member told me a long story about how a particular faculty job at a prestigious institution went to an under-qualified woman rather than the highly talented man the job "was intended for," indeed, that a Dean had insisted on adding this woman to the short list. This was horrible to hear. No one wants to be told that she will get a job unfairly. But I had the sense to ask a few questions: Who was the woman? He didn't know. What did she work on? He didn't know. I started to have my suspicions that perhaps he didn't know the story as well as he pretended. Years later, I talked to someone who had been on the actual search committee, who told me that the woman gave a spectacular talk and blew them all away, which is why she was hired.

This episode started me down the path of gender activism. I watched and listened as women around me were overlooked, undervalued, mistreated, or harassed. (Strangely, I never noticed this behavior directed at me, although in retrospect that was certainly the case.) Women's suggestions were routinely ignored, only to be resurrected and appreciated when later raised by a man. One colleague was constantly second-guessed, unlike any of her male counterparts, and when she pointed this out, was told she was depressed and should get professional help. Another woman told me it had become routine for her to cry while driving home from work. The idea that women were somehow privileged in the scientific world simply didn't hold water.

I wish I had been able to ignore all this but after a while, noticing the scarcity of women being hired into faculty jobs and, frankly, getting precious little positive feedback, I began to believe that I wasn't good enough. When I expressed ambition, I was put down. When I asked a potential employer to match a far better faculty offer elsewhere, he declined to change the salary or the position he originally offered; when I said that seemed unreasonable, he replied, "Maybe you're not as good as you think you are." (I went off to cry for a few hours. Then I accepted his too-low offer because there were other constraints.) When I suggested I was ready to be tenured, I was told, "Be patient, Meg, it's too early for you." (I was tenured about a year later.) When I mentioned I was interested in a high-level national committee, the response was, "Isn't that a bit ambitious, Meg?" (I was appointed to that committee within a year of that comment.) When I expressed interest in a promotion, my then-boss said, "You're not a leader, no one would follow you." Well, not with that kind of support, that's for sure.

Throughout this time, I was breaking new ground in my research, publishing important, highly cited papers, and raising substantial funding through grants. Fighting past the discouragement and discrimination, I built a successful career, and I now have a very satisfying position and plenty of recognition. I can still be overly self-critical, but the Doubting Thomas in my head is a lot quieter now. Most of the time, I think I am really great at my work. But believe me, I still "get" the Cinderella thing.

Reflections and Recommendations

One of the hardest parts of getting to where I am in my career has been overcoming the insidious training I had, as does any girl in our society, to be a quintessential female: to be self-effacing, to avoid "bragging," to support others even at the expense of foregoing appropriate credit oneself – all wonderful, polite things, but very much at odds with the dominant scientific culture today, at least in the U.S. Learning about a successful woman, especially in a field where she had to fight for her right to a seat at the table, was the most incredible boost. Finding a few women ahead of me and more in my peer group and even more coming up behind has been critical to my staying in astronomy. I hope that young girls considering careers in science can find the same source of strength and inspiration in my story that I took from those women who went before (in much tougher times). And that they will heed some of this advice:

All children are born scientists because they are naturally curious. Science is a very satisfying career because it engages the mind. The most interesting part of my job is learning new things and making progress toward understanding our universe. Each day, the new thoughts, the new ideas, and the exercising of one's brain make it all worthwhile.

It's really important to have a life besides a career. I like science a lot, but it isn't the most important thing; my family is. I have a great husband and two adorable daughters, Amelia and Sophia, and I love them more than it is possible to explain. It's tough to manage the family plus job sometimes – and, by the way, I think it's just as tough, or maybe tougher, for women who stay home and raise their children without much help. Most women with small children also work outside the home, but their jobs aren't as satisfying, flexible, and rewarding as mine.

The most important advice for young women who want both a career and a family is: marry the right person! I am really surprised sometimes when I hear intelligent young women agreeing to shoulder the greater part of the work of raising their children, agreeing to subordinate their careers and aspirations to their partner's, for no reason other than that is the way it is always done or that is the way their husbands and boyfriends have assumed it will be. I hope young girls grow up valuing their dreams and their futures as much as young boys do. The main reason I could have both a career and a family is because my husband, who is also an astrophysicist, is an equal partner in our marriage. He doesn't "help" – we share. We made it equal, from start to finish – Amelia and Sophia even carry both our last names as their middle and last names, but in alternate order.

Be proud of your ambitions – not ashamed of them. There is nothing wrong with being the smartest kid in the class. There *is* something wrong with doing less for the world than you are capable of. Our world needs every contribution, from everyone who has something to give.

LAKEISHA MARIA
HOGUE WALKER

Bona Fides

Ms. Lakeisha Maria Hogue Walker received her Bachelor's degree in Materials Science and Engineering from the University of Illinois at Urbana-Champaign (UIUC) in 2001. With support from her family, a strong will, hard work, and determination Lakeisha has overcome several socioeconomic hurdles and gone on to have a successful career in the nuclear weapons complex (with Babcock and Wilcox Technologies, Inc). Seeking new challenges, Lakeisha started a second career as a scientific associate at Oak Ridge National Laboratory (ORNL) in 2004, performing neutron scattering studies. She was nominated for ORNL's "Outstanding Female in Research" award in 2007, received ORNL's "Outstanding Mentor Award" in 2008, and recently was accepted into the "Teach TN" program, allowing her to share her knowledge with the next generation of physicists.

A Precious Life

Statistics say that children of teen moms do worse in school than those born to older parents – with half of these children failing at least one grade. Those same children are less likely to finish high school and tend to perform more poorly on standardized. tests. Many children born to teen moms have behavioral problems, juvenile delinquency, and conflict with authority . . .

I was a statistical anomaly from the start. Despite being born to a 14-year-old mother, I always excelled in school – I was ahead of learning curves in reading, writing, and mathematics. I also excelled in social interactions; my mother tells stories of my childhood in which friends and family would beg to babysit me and take me places. It seemed wherever I went I was very well behaved, and I astounded everyone with my clear speech (in fact, my nickname was Ms. Magoo, after the cartoon character Mr. Magoo, because of my big forehead, small stature, and mature speech). My mom instilled in my sister and me the value of education, and we spent countless hours at the public library, on discovery field trips, and just talking about things.

I never had a desire for math, science, or engineering in primary school. They seemed too far-fetched and constraining, and I never saw them as creative – something I wanted to be. I knew that I would be a dancer or writer, although when asked I said I wanted to be a lawyer or doctor (even though I had no desire to be those either – they just sounded like good choices). In secondary school I became fascinated with materials used in weaponry. I had questions like, "How could a gun be made of something undetectable by metal detectors?" I was also fascinated by the mystery and intrigue of assassins and espionage, which led to a pre-occupation with government conspiracies and FBI/CIA agents. I wanted to work for the government in that capacity. I wanted to make special "Mission Impossible" style weapons.

Change of Plans

With this newfound inclination towards design and engineering I realized I needed to enter into college with strong math and science skills. So I asked my mom if I could apply for Austin's math and science magnet school at LBJ High School. By junior year I was moving full speed ahead towards a government career in engineering. My mom was an IRS agent at the time, and her work always seemed so official and important. I wanted to be like her but with my own twist.

All was going according to plan until I got tangled up with a young man for whom school was an afterthought. I became distracted, and much to my mother's dismay, I walked across my high school graduation stage the summer of 1995 five months pregnant. Now I *was* a statistical norm. I was embarrassed, and my dreams of working with the FBI were shattered. My mom, however, encouraged me to keep moving forward. After moving to St. Louis with my family I went back to school part-time the fall semester after graduation. I was still pregnant when I began my first class and required special accommodations because I could not fit in the student desks. After my daughter Amana was born, my mom and younger sister watched her in the evenings while I attended school part-time. Amana was, in a lot of ways, easy to raise, and this was such a blessing. She learned to nurse without any issues and slept through the night almost immediately, which helped me realize that I could be more than a high school graduate.

In the summer of 1997 I moved to Champaign, IL and began classes at Parkland (community) College. This was a great move for me because Parkland offered an Engineering Science program, which provided a seamless transition to the four-year engineering programs at the University of Illinois Urbana-Champaign (UIUC). Community college was AWESOME! Classes were small, instructors were caring, and the small campus easily accommodated non-traditional students like myself. I treated the school day like my full-time job. I studied in between classes and completed everything between 8 AM and 4 PM each day, which allowed me to focus on Amana in the evenings. Government welfare programs helped with groceries, rent, and utilities, and I had a work-study job in the college's tutor lab. Things were hard, but they were manageable, and I was encouraged to keep moving forward.

Moving Forward

The summer of 1998 was busier than ever as I planned my wedding. My fiancé, a former Marine and hard-working man, was my dream come true: a man I believed would be a wonderful father for my daughter, as he was with his own infant daughter. With marriage we became a ready-made family of four, and soon after my home responsibilities grew. My desire to finish school and become an engineer transitioned from being driven by exploration to being driven by the desire for a high-paying job. I felt an overwhelming pressure to provide for my family, and I believed that a degree in engineering could help me accomplish this without the additional requirement of graduate school. At UIUC I worked as a research assistant alongside one of my advisor's graduate students. Classes within my specialty, Polymer Engineering, were very small, and I was *always* the only female and often the only brown-skinned person.

My demographic and home-life differences made finding students to study with virtually impossible. While I wanted to do it all during the day, my comrades preferred to do their studying in the evenings, on campus, which was a good distance from my home. I also had a family who I wanted to spend my evenings with. I did join the late-night crunches for major exams/finals, but unlike at community college, I was left to study alone most days. As a result my studies declined and so did my grades. By my final year in college I was exhausted and ready for graduation.

In the summer of 2001 I graduated with my Bachelor of Science degree in Materials Science and Engineering from the University of Illinois (UI). That same year U.S. News and World Report ranked UI's undergraduate Materials Science and Engineering program #1 in the nation. This was a blessing for me because although my grades were not the best, the name of my university carried weight within the scientific and engineering community. I had an engineering job lined up with weapons manufacturer Lockheed Martin before the spring of my graduating year. Out of all my job offers, the nuclear weapons complex seemed the most appealing and challenging. My family was apprehensive about the move from Illinois to Tennessee, but we were excited for the change and the increased income.

On The Grind

By the time I reported to work at Y-12, Lockheed Martin was no longer the managing agency, and the facility was being managed by Babcock and Wilcox Technologies, Inc. (BWXT). The level of security and the process of obtaining a security clearance blew me away. I had never been asked to chronicle my life in such detail! One month later, on September 11, 2001, I was in a new employee training session when the announcement about the Twin Towers was made and the facility was placed on lockdown. It was very humbling knowing that it could have been us that was targeted. From then on the security requirements became ridiculous, and what was once exciting had become tedious. To top it all off, my exciting, secretive job required more paper pushing than I ever dreamed imaginable. My engineering classes hadn't prepared me for writing safety bases and procedures, nor for documenting every modification. Technical writing was now 90% of my job, and it wasn't very satisfying. Within 3 years I was eyeing jobs at the "country club" over the ridge – Oak Ridge National Lab (ORNL) – and imagining the fulfillment of a research and development job.

This dream was realized when I became a Scientific Associate at The Spallation Neutron Source (SNS) at ORNL. State of the art in every way, SNS was a place of ideas, growth and technology. There was so much to learn and do; new opportunities presented themselves at every turn. Today I have found my focus with the Neutron Imaging Instrument at the High Flux Isotope Reactor (HFIR) at ORNL, and I can say with certainty that push after push has brought me into new and better experiences. I give honor to Yeshua for my growth and opportunities. I am now surrounded by a wonderful group of people, and I am finding that when the team is right the rest will fall into place.

ALICE E. WHITE

Bona Fides

Dr. Alice E. White became Chief Scientist of Bell Labs in May 2011. In this role, she is responsible for the long-term research strategy, university partnerships, and the Bell Labs Technical Journal, as well as maintaining technical excellence through technology- and science-recognition programs. She has a Ph.D. in Physics from Harvard University and a broad technical background in experimental solid-state physics and fabrication of optical components. Since 1989, she has held various leadership positions at Bell Labs including Director of Materials Physics Research, Director of Integrated Photonics Research, VP of the Physical Technologies Research Center, and location leader for Bell Labs North America. In 1991, she received the Maria Goeppert-Mayer Award of the American Physical Society for her work on compound formation using ion implantation. She was named a Bell Labs Fellow in 2001 for her work in "developing and applying novel integrated photonic device technologies in advanced optical networks." With over 125 publications, she is a fellow of the American Physical Society, the IEEE Photonics Society, and the Optical Society of America.

First Discoveries

One of my earliest memories is of my dad arriving home from

258

Bell Labs, where he'd been working on a Saturday, and announcing that he and his colleagues had gotten a laser to emit light. I remember his excitement and my mom's interest. She asked, "How will it be used?" He laughed and answered that it was "just a neat thing." Then he suggested that we go out to dinner – a rarity for our family – so I knew something important had happened. The laser that he referred to was the visible helium-neon (HeNe) laser, the red line that turned out to have a myriad of practical uses. That lesson – that science was fun and could also be useful – has stuck with me through the years.

My mother was the true pioneer, however. She had gone to Middlebury College to major in French but took a physics course to meet a general education requirement and fell in love with the subject. My path at Middlebury was similar – I entered as a chemistry major, having endured a terrible physics teacher in high school. I ended up taking Physics 101 at the same time as organic chemistry – also to meet a distribution requirement – and realized that physics was much more to my liking as it used more math (which I love) and less memorization (for which I have no patience).

Middlebury's science and math programs were outstanding. In addition to wonderful teaching, I had the opportunity to spend a January term at the Massachusetts Institute of Technology (MIT), where I had my first personal experience of discovery. Analyzing data from the Uhuru satellite (done back then by submitting a stack of commands on IBM cards to the mainframe computer), I discovered a large change in the X-ray flux coming from the galaxy Centaurus A, my assigned object. Prof. Frank Winkler and I checked and re-checked the data, pored over the previous literature, came up with a plausible explanation, and decided to publish what was my first paper.

From Bell Labs to Harvard

That summer, I was accepted into the newly established

Summer Research Program for Women and Underrepresented Minorities at Bell Labs in Murray Hill, NJ. A cadre of enlightened Bell Labs scientists, including Jim West, Lou Lanzerotti, and Bob Dynes, had decided to do something about the woefully inadequate pipeline of women and minority Ph.D. candidates for researcher positions. I worked with Pat Cladis, researching liquid crystals. Pat was a fearless experimentalist with a gung-ho approach: in three words, "Let's try it!" (an approach I've tried to exemplify in my own career). At one point during that summer, she acquired an especially large and pure sample of a nematic liquid crystal and wondered if she could orient a macroscopic quantity. The experiment involved a 5-gallon circulating oil bath (to maintain the temperature). To observe progress, the oil needed to be index-matched to the glass, i.e., microscope immersion oil. The only source of oil we could find was in 2-oz jars, so we spent a couple of hours opening jars and pouring the oil into the bath. Five days later, the sample had aligned and was completely transparent except for a disclination down the middle. Probing the structure with a HeNe laser, we discovered that the disclination acted as a polarizing beam splitter. That weekend, one of the hoses popped off the oil bath, spraying oil over the entire lab – even that mess didn't faze Pat. The rest of the summer, oil was seeping out from under the lab benches. Luckily, we had gotten all the data we needed.

When I returned to college for my senior year, my Middlebury professors urged me to apply to graduate school, something that hadn't really been on my radar screen. With their encouragement, I applied to several schools, eventually choosing Harvard because I thought, wrongly, that it would be an environment similar to what I had experienced at Middlebury. I was awarded a Bell Labs Graduate Research Program for Women Fellowship – those same enlightened researchers behind the summer research program also realized that financial and mentoring support were important ingredients to keep women and minorities from dropping out of grad school. I was very lucky to have Doug Osheroff as my Bell Labs mentor for the fellowship program. He was disappointed that I did not choose to attend his alma mater, Cornell, but he still gave me unswerving support, an independent perspective, and a boost of energy each time I visited Bell Labs during my six years in grad school.

Unlike my Middlebury experience, the other (all male) students in my classes at Harvard were not supportive. One student suggested that it was too bad that "women's liberation forced you to go to grad school when you'd much rather be home raising a family." Things improved dramatically, though, when I was accepted into Prof. Michael Tinkham's research group. I was his first ever woman student – the apocryphal story going around the group was that, since he had accepted a man with a ponytail, he guessed he could handle a woman. Although we each had our own individual project, Prof. Tinkham had lunch with us every day and held weekly meetings where students shared their work with the group. We would pepper each other with questions – great training for later talks at professional meetings and conferences.

Taking an Idea and Running with it

Although the postdocs and students in the Tinkham group were very helpful, I quickly realized that the project I was supposed to do – studying the transport properties of superconducting tin microbridges – was beyond my skill and patience level. To make the microbridges, I was evaporating tin films on glass substrates with a hand-drawn scratch and creating the bridge by bringing down a diamond edge on the film over the scratch. I needed to make submicrometer bridges, but the microscope in the lab could only resolve a micrometer, so I was flying in the dark. Previous students had succeeded only with hundreds of attempts. That fall, Prof. Tinkham went on leave, and I signed up for a course in submicrometer fabrication taught by Prof. Hank Smith at MIT. I quickly realized that the techniques his group was pioneering at MIT's Lincoln Labs were exactly what I needed. I jumped at the chance to join his group and make my samples under Dale Flanders' guidance. This was an important turning point for me and another important lesson – grab opportunities to try something new when they arise. The techniques that I learned working with Dale I have used again and again. I returned to Harvard with 10 nanometer normal metal wires that I used to study the then-hot topic of Anderson localization, instead of my original problem. To Prof. Tinkham's credit, when he came back from leave, he embraced the change in direction and provided the funding for me to build the first clean room at Harvard (now there is a multi-story Nanotech building next door!) When I finally got data, he delighted in trying to understand what it meant. Our publication came out in Physical Review Letters a few months later.

When it came time to look for a job, I didn't even consider academic positions, gravitating toward the big industrial research labs instead. I still remember one of the questions posed during my interview seminar at Bell Labs – someone asked me if I had done tunneling to measure the density of states in my 1-D wires. Another researcher quickly asserted that that was impossible, but I heard myself saying that I thought it could be done. I was offered a fulltime position doing applied materials research but instead accepted a postdoctoral appointment in fundamental physics, realizing that I could always re-evaluate at the end of the two-year appointment.

I hit the ground running at Bell Labs, which has a wonderfully collaborative environment. I'm sure that as a woman I was met with skepticism, but the other researchers soon learned about my fabrication skills and low-temperature experience and immediately realized what I could do to help them further their own projects. Once I developed credibility, I became accepted as a full-fledged member of the technical staff.

I found little evidence of hierarchy at Bell. In those early days,

262

I had the opportunity to interact with Phil Anderson, who had already earned a Nobel Prize, and Horst Stormer, who would later do so. Much like Prof. Tinkham, Phil Anderson loved to look at data. I remember when I finally was able to make tunneling measurements on some of those 1-D wires, the curves had a surprising discontinuity. Phil stopped by the screen room to watch the data arrive and challenged whether or not I had zeroed the phase on the lock-in – a very savvy question for a theorist! Horst used to barrel down the hall from his office in his enthusiasm to get into his lab; I learned to look both ways before crossing the corridor. In the lunchtime discussions, regular Journal Club sessions, and endless hallway conversations, I continued to learn and explore.

In 1982, Judge Green handed down the consent decree breaking up the Bell System and turning AT&T (then long-distance only) into a commercial entity. As my postdoc was ending, ten percent of the Bell Labs research staff left to form Bellcore, the research lab for the local telephone companies – the "Baby Bells" – which were still a regulated monopoly. It was a time of great uncertainty, and I interviewed for several junior faculty jobs, a process that was frankly humiliating. I had been treated as an equal for two years at Bell, but it was clear that in many cases, I was the token female candidate at the universities. At one university, I was shown the door after my seminar. Back at Bell, I had the opportunity to take a permanent staff position either continuing the research I had started or doing something completely different, and I chose the new direction.

Bell Labs: The Idea Factory[10]

My new field of research was ion implantation in silicon, a

---

[10] Subtitle inspired by The Idea Factory: Bell Labs and the Great Age of American Innovation. Jon Gertner, Penguin Press HC, 2012.

technique that was being used commercially at low doses to dope silicon for electronic integrated circuits. I decided to look at high-dose ion implantation instead. Here I learned firsthand the benefits of interdisciplinary collaboration and switching fields. I decided to try and create a buried silicide layer in the silicon wafer by implanting cobalt at high temperatures. The ion implantation gurus told me that ion-beam-induced sputtering would make this impossible, but I insisted that we try it anyway. Much to their (and my) surprise, we created a single-crystal layer of $CoSi_2$ buried below a surface layer of crystalline silicon, something that had never been done by any other technique. Calling the process "Mesotaxy" (by analogy with Epitaxy), my collaborators and I spent several fruitful years exploring this field and the applications of the buried metal layers. In 1987, the American Physical Society (APS) awarded me the Maria Goeppert-Mayer Award for my work in Mesotaxy.

During my 30 years at Bell, I've had marvelous opportunities, moving into new areas and management roles (reluctantly at first). I learned to appreciate the hard work that goes into turning a research concept into a product and, as Bell Labs has become more global, how to navigate different cultures. Along the way, I have worked with the most amazing people, all of whom have been generous in sharing their knowledge and experience. I met my husband, Don Monroe, at the Labs, a convenient solution to the "two-body problem." His support and understanding of the demands of my job have been absolutely key to my success. Together, with support from my parents, we have raised two talented and independent daughters. As they head off to college, I'm contemplating the next chapter of my life as well.

ABOUT THE EDITORS

Rhiannon Meharchand is a Director's Postdoctoral Fellow at Los Alamos National Laboratory. She received her Ph.D. in Physics from Michigan State University (MSU) in 2011, her Master's degree in Physics from MSU in 2008, and her Bachelor's degree from Florida State University in 2006. An experimental nuclear physicist by training, her dissertation research used charge-exchange reactions to probe the structure of unstable nuclei. Her postdoctoral research involves developing and testing a time projection chamber for use in high-precision fission cross-section measurements.

Emma Ideal is a fourth-year Ph.D. candidate in Physics at Yale University. She received her M.S. and M.Phil. degrees in Physics at Yale in 2010 and 2012, respectively, and her Bachelor's degree from the University of California, Los Angeles in 2009. She is a National Science Foundation Graduate Fellow. As a member of the ATLAS experiment at CERN in Geneva, Switzerland, her dissertation research focuses on a search for the Higgs Boson, the most elusive of the fundamental particles predicted by the Standard Model.

Both editors have been actively involved in science outreach and diversity efforts and have received awards and recognition for both scientific and service-related achievements, including the Association for Women in Science Luise Meyer-Schutzmeister Memorial Award (Meharchand, 2010) and Kirsten R. Lorentzen Award (Ideal, 2012).

32072954R00156

Made in the USA
San Bernardino, CA
26 March 2016